Dispersive Soils

To Dr WW Emerson (formerly of the CSIRO Division of Soils), whose pioneering investigations into clay dispersion from soil aggregates inspired generations of scientists – including us – to pursue research on dispersive soils.

Dispersive Soils

Processes, Impact, and Management

Pichu Rengasamy and Ehsan Tavakkoli

PUBLISHING

A catalogue record for this book is available from the National Library of Australia

ISBN: 9781486319794 (pbk)
ISBN: 9781486319800 (epdf)
ISBN: 9781486319817 (epub)

How to cite:
Rengasamy P, Tavakkoli E (2026) *Dispersive Soils: Processes, Impact, and Management*. CSIRO Publishing, Melbourne.

Published by:

CSIRO Publishing
36 Gardiner Road, Clayton VIC 3168
Private Bag 10, Clayton South VIC 3169
Australia

Telephone: [+613] 9545 8555
Email: csiropublishing@csiro.au
Website: www.publishing.csiro.au
Sign up to our email alerts: publishing.csiro.au/earlyalert

Front cover: (Top) Farmer standing on a dispersive soil (photo by Pichu Rengasamy). (Bottom, left to right) Erosion from dispersive clay in a riparian landscape (photo by Ehsan Tavakkoli); Gully erosion in a dispersive soil in Tasmania (photo by Dr Richard Doyle, University of Tasmania); Piping and initiation of sinkholes on a hillside in Tasmania (photo by Dr Richard Doyle, University of Tasmania)

Edited by Natalie Korszniak
Cover design by Cath Pirret Design
Typeset by Envisage Information Technology
Printed in Australia by Fraser & Jenkinson Pty Ltd

CSIRO acknowledges the Traditional Owners of the lands that we live and work on across Australia and pays its respect to Elders past and present. CSIRO recognises that Aboriginal and Torres Strait Islander peoples have made and will continue to make extraordinary contributions to all aspects of Australian life including culture, economy and science. The use of Western science in this publication should not be interpreted as diminishing the knowledge of plants, animals and environment from Indigenous ecological knowledge systems.

The paper this book is printed on is in accordance with the standards of the Forest Stewardship Council® and other controlled material. The FSC® promotes environmentally responsible, socially beneficial and economically viable management of the world's forests.

Dec25_01

Contents

Preface

IT IS EARLY MORNING IN the wheat–sheep belt of south-eastern Australia. The sun hangs low, casting long shadows across a paddock that, at first glance, appears perfectly ready for planting. But a few steps into the field tell another story. The topsoil is crusted, brittle underfoot. Where a boot presses down, fine cracks radiate outward, and a thin veil of dust lifts into the cool air. The farmer gestures towards shallow rills where rainwater has carved channels, carrying away not only sediment but also the promise of yield.

'It's the same every year,' he says quietly. 'No matter what we try, the soil just doesn't hold together.'

Scenes like this are neither rare nor recent. Across Australia – and in many other parts of the world – dispersive soils quietly erode agricultural productivity, degrade water quality, and compromise infrastructure. From the croplands of the Murray–Darling Basin to the rangelands of North America and the alluvial plains of South Asia, their impact transcends borders and climates. At the heart of the problem lies the behaviour of clay particles. When wetted, they lose cohesion and disperse into fine suspensions that clog pores, seal soil surfaces, and collapse structure. The consequences cascade: infiltration slows, roots struggle, nutrients leach, and erosion accelerates. What appears as a local surface crust or a failed crop is, in fact, the visible symptom of a deeper, systemic fragility.

A PROBLEM AS OLD AS AGRICULTURE

The challenges posed by sodic and dispersive soils are as old as agriculture itself. For millennia, salinisation and structural decline have haunted the landscapes that once sustained thriving communities – and, in some cases, brought them to collapse. In southern Mesopotamia, across the fertile plains of the Tigris–Euphrates valley, rising soil salinity and structural degradation slowly eroded the productive base of societies that had flourished for centuries, contributing to their eventual demise (Jacobsen T, Adams RM (1958) Salt and silt in Ancient Mesopotamian agriculture: progressive changes in soil salinity and sedimentation contributed to the breakup of past civilizations. *Science* **128**, 1251–1258). Similar patterns emerged in other corners of the ancient world, where the interplay of irrigation, salinity, and soil chemistry steadily undermined farming resilience.

In the modern era, systematic study of salinity and sodicity accelerated from the early 20th century, with pioneering work at the United States Salinity Laboratory. Salinity was recognised not only for its direct effects on plants – through osmotic stress and ion toxicity – but also for its indirect impact on soil properties and erosion risk. Central to these indirect effects, as understood at the time, was the phenomenon of clay dispersion – typically attributed to high levels of exchangeable sodium – which destabilised aggregates and degraded soil structure.

THE AUSTRALIAN SCIENTIFIC LINEAGE

In Australia, the late Dr WW Emerson of the CSIRO Division of Soils made seminal contributions to understanding dispersion. His now-famous *crumb test* provided a simple yet powerful way to identify dispersive soils in the field. Emerson's work inspired generations of soil scientists, both in Australia and abroad, to probe the mechanisms and management of sodic soils.

The U.S. definition of sodic soils, based on an Exchangeable Sodium Percentage (ESP) greater than 15, was adopted internationally. However, Australian research – informed by local conditions – demonstrated that dispersion could occur at much lower ESP values. This led to the Australian definition of sodic soils at ESP ≥6, reflecting a more realistic threshold for structural risk in many environments.

FROM IONICITY TO CROSSC: CONCEPTUAL BREAKTHROUGHS

Since 1975, the senior author (PR) has served the field of soil science in Australia with exceptional distinction, advancing understanding of the chemical physics of soil – how its components and their interactions govern structure, stability, and productivity. While working on irrigated soils in Victoria, he investigated the factors influencing clay dispersivity, from organic matter and pH to clay mineralogy and electrolyte concentration.

A pivotal insight came during laboratory experiments: clay dispersion occurs when polar water molecules interact with the net ionic charge of soil particles, overriding previous explanations involving rainfall impact energy or air expansion within aggregates. This was confirmed in collaborative experiments with Professor Malcolm Sumner at the University of Georgia, where rainfall simulations using both water and diesel (a non-polar liquid) confirmed the role of water–clay interactions. From this work emerged the concept of *ionicity of clay–cation bonds* and the formulation of *ionicity indices* for common cations.

Later, Dr Alla Marchuk's PhD research confirmed that dispersion was closely linked to the ionicity index, across a suite of homo-ionic clays containing Li, Na,

K, Mg, Ca, and Ba. These findings opened the way to a new, more mechanistic understanding of sodicity, dispersion, and their management.

COLLABORATION AND NEW INDICES

In 2007, the junior author (ET) began his PhD research with Pichu as co-supervisor. This initial mentorship developed into a sustained research partnership spanning nearly two decades, blending theoretical insights with applied, industry-focused projects. Together with Professor Glenn McDonald, they undertook extensive field investigations into subsoil constraints across south-eastern Australia.

Analysing hundreds of soil samples for exchangeable and soluble cations, dispersivity, and related properties, the team validated the concept of *net dispersive charge* – the balance between dispersive and flocculating forces in a soil's cation suite. This parameter proved to be strongly correlated with actual field dispersivity. Building on this, they jointly developed the Cationic Charge Ratio for Soil Structural Stability (CROSSc) index, integrating dispersive and flocculating charges and their interactions. CROSSc now offers a more robust tool than SAR or CROSS for evaluating irrigation water quality and predicting dispersion risk.

SODIC VS DISPERSIVE: A SHIFT IN PERSPECTIVE

A central theme of this book is a shift from the input-based concept of the 'sodic soil' to the outcome-based concept of the 'dispersive soil'. While the two are related, they are not interchangeable. Sodicity describes a chemical state – high sodium on the soil's exchange complex – that may predispose the soil to structural problems. Dispersiveness describes actual behaviour: whether the soil holds together or falls apart when wet.

A soil may be sodic yet stable, or dispersive without being sodic, depending on factors such as electrolyte balance, clay mineralogy, and organic matter content. This distinction is not academic – it is central to diagnosis and management. Decades of research by many scientists and our own findings have shown that reliance on a single sodium threshold can be misleading. What matters to farmers, engineers, and environmental managers worldwide is whether a soil retains its structure under wet conditions. By focusing on 'dispersive soils', we address the materials most prone to erosion, tunnelling, gully formation, and crop failure in relation to the balance between dispersive and flocculating charges of cations – whatever their ESP.

FROM SCIENCE TO STEWARDSHIP

The science in this book is not an abstract exercise. In agriculture, dispersive soils limit yield potential; in engineering, they threaten the stability of embankments and earthworks; in environmental systems, they drive sediment and nutrient loads into waterways – for example, from Australia's Murray–Darling Basin, or to Asia's Indus plains, or to North America's rangelands.

Our work spans conceptual advances and practical interventions – targeted amendment placement, engineered organic–inorganic blends, drainage design, and water quality management – all framed within a systems approach. The concepts of net dispersive charge and CROSSc are equally applicable to crop management, catchment planning, salinity control, safeguards in building and infrastructure projects, and erosion mitigation.

THE PURPOSE OF THIS BOOK

Dispersive Soils is both a synthesis and a call to action. It synthesises the global and Australian history of dispersive soil research, from ancient agricultural collapse to modern field trials. It challenges the field to move beyond simplistic thresholds and single-ion models, towards integrated frameworks that capture real-world complexity.

Our aim is to resolve the long-standing conundrum of sodic/dispersive soils – to clarify their chemical dynamics, connect these to their physical expressions, and provide practical, sustainable remediation strategies. We hope this book will serve scientists, engineers, land managers, and policymakers worldwide, fostering not only innovative thinking and collaboration but also decisive, effective action.

LOOKING AHEAD

The future of dispersive soils research and management will be shaped by:

- Precision diagnostics – remote sensing, proximal sensing, and AI-driven mapping of dispersion risk at multiple scales.
- Engineered amendment systems – tailored blends optimised for specific soils and management goals.
- Soil–carbon–water integration – recognising dispersive soils as both a challenge and an opportunity in climate-smart agriculture and carbon sequestration.

- Catchment-scale planning – linking paddock interventions to downstream water quality outcomes.
- Socio-economic adoption frameworks – ensuring solutions are practical, profitable, and culturally relevant.

CLOSING REMARKS

This book is the culmination of decades of research, fieldwork, and collaboration. We are indebted to the farmers, colleagues, and students who have shared their observations, data, and questions – and to the institutions that supported our work.

From the irrigation systems of ancient Mesopotamia to today's satellite-guided precision agriculture, the story of dispersive soils is one of persistence, challenge, and adaptation. Our hope is that the concepts, evidence, and strategies presented here will help transform dispersive soils from a constraint endured into a challenge that can be overcome.

The crust on that early morning paddock may still be there today, but the knowledge to change it – and the willingness to act – have never been stronger.

Acknowledgements

WE WISH TO EXPRESS OUR deepest gratitude to the many colleagues, students, and collaborators whose insights, encouragement, and generosity have shaped and enriched the content and scope of this work. Over the course of its development, we have been privileged to draw upon the knowledge, creativity, and critical feedback of an exceptional community of peers and students, whose contributions have left a lasting imprint on the pages that follow.

We acknowledge with sincere appreciation the expertise and shared resources provided by Mr David Armstrong, Dr Roger Armstrong, Dr Ed Barrett-Lennard, Dr David Brautigan, Dr Mostafa Chorom, Dr Jock Churchman, Dr Yunying Fang, Dr Richard Greene, Mr David Hall, Mr Shane Hildebrand, Mrs Yan Jia, Professor Enzo Lombi, Dr Alla Marchuk, Professor Glenn McDonald, Professor Michael McLaughlin, Dr Wayne Pitt, Professor Malcolm Sumner, Dr Lukas Van Zwieten, and Dr Le Yu. Their willingness to share ideas, data, and experience has greatly deepened the scientific foundations of this book. We also gratefully acknowledge the contributions of technical officers, laboratory staff, and research assistants whose expertise in field operations, sample processing, and data management ensured the rigour and reliability of the findings presented here.

We are equally indebted to the many farmers across Australia who have collaborated with us over the years – hosting field trials, facilitating soil sampling, and generously sharing their observations, experience, and perspectives on the real-world challenges facing agriculture. Their openness and engagement have been instrumental in shaping our research to address practical needs, ensuring that the outcomes remain relevant and valuable for end users.

We are especially grateful to Professor Richard Doyle and Mr Jim Kelly for their invaluable assistance in developing several of the figures presented herein. Their skill in distilling complex scientific concepts into clear, visually engaging forms has enhanced the accessibility and impact of the work.

Our thanks also go to Professor C Tang, whose introduction to the CSIRO Publishing team set in motion a collaboration defined by professionalism, creativity, and care. We wish to particularly recognise Dr George Knott, Mark Hamilton, and Tracey Kudis from CSIRO Publishing, and Dr Natalie Korszniak, for their meticulous attention to detail, patient guidance, and steadfast commitment to excellence throughout every stage of the publication process.

We are indebted to the funding bodies and research institutions that have supported our work over the past two decades – including the Grains Research and Development Corporation, Cooperative Research Centre for High Performance Soils, ANSTO, and the Department of Agriculture, Fisheries and Forestry – as well as the University of Adelaide, CSIRO, the Australian Synchrotron, the University of South Australia, the University of New South Wales, and the NSW Department of Primary Industries. Their sustained investment in agricultural research has provided the essential resources and infrastructure that underpin much of the knowledge presented in this book.

Pichu Rengasamy wishes to acknowledge the unwavering support and encouragement of his wife, Veda, and his children, Raja, Jaya, and Sundara, whose understanding and patience have been a constant source of motivation throughout the years of work that culminated in this volume.

Ehsan Tavakkoli wishes to acknowledge the unwavering love and support of his parents, Farideh and Kazem, his sister, Professor Mitra Tavakkoli, his brother-in-law, Dr Hassan Fadavi, and his niece, Baran, whose presence and encouragement have been a continual source of strength and inspiration.

Finally, we extend our collective thanks to the broader scientific community working in the field of soil science. It is upon their collective achievements, curiosity, and commitment to advancing knowledge that the foundations of this book firmly reside.

Pichu Rengasamy
Visiting Research Fellow, Adelaide University
Adjunct Research Fellow, La Trobe University

Ehsan Tavakkoli
Associate Professor, Adelaide University

1

Understanding dispersive soils: Moving beyond sodicity

1.1 OBJECTIVES OF THE BOOK: MOVING FORWARD

Dispersive soils, those whose clay particles readily disperse in water, pose significant challenges in agriculture, engineering, and environmental management. They can lead to problems such as soil structure collapse, erosion, sedimentation, and reduced crop productivity. This book brings together current scientific understanding of dispersive soils to help academic soil scientists, postgraduate students, policymakers, and on-ground practitioners move forward with a clearer, more effective framework for identifying and managing these problematic soils. In particular, we aim to advance beyond traditional notions of 'sodic soils' (defined solely by sodium content; Naidu *et al.* 1995) towards a more practical emphasis on clay behaviour (dispersion). The overarching objectives of the book are outlined below:

- clarify key concepts: provide clear definitions for sodic soils (based on soil chemistry) and dispersive soils (based on soil physical behaviour) and explain why these terms are not interchangeable. This involves dissecting the relationship between exchangeable sodium levels and clay dispersion in soils
- identify limitations of current criteria: critically examine the limitations of existing sodicity indices (such as the exchangeable sodium percentage (ESP) and sodium adsorption ratio (SAR)) in predicting soil dispersion. We highlight that no single ESP threshold universally distinguishes dispersive from non-dispersive soils; for example, some soils with very low ESP can still disperse, whereas others with high ESP may remain stable (Rengasamy *et al.* 2016)
- promote a behaviour-based approach: advocate for a shift in diagnostic focus from purely chemical metrics (ESP/SAR) to direct measures of clay behaviour in water. By emphasising tests of aggregate stability and dispersion, the book underscores clay dispersion as the key criterion for identifying problematic soils. This approach aligns with recent research showing that combining chemical indicators with dispersion tests improves predictive accuracy

(Rengasamy and Marchuk 2011; Rengasamy *et al.* 2016; McKenzie and Bryce 2023)

- guide practical management: provide science-backed strategies for managing dispersive soils, moving beyond one-size-fits-all remedies. The book connects theory to practice (e.g. explaining why amendments like gypsum work by altering clay behaviour) so that policymakers and practitioners can make informed decisions
- bridge disciplines and update knowledge: synthesise insights from soil chemistry, physics, and field case studies (including recent peer-reviewed findings up to 2025) into an accessible narrative. This ensures that advanced students and professionals are equipped with the latest understanding and tools to address dispersive soils in their work.

By moving forward with these objectives, the book sets the stage for an updated paradigm in soil science. The following sections introduce the core concepts of sodicity versus dispersivity, and why a renewed focus on dispersive soils is warranted.

1.2 DISPERSIVE SOILS AND SODIC SOILS

The terms 'sodic soil' and 'dispersive soil' have often been used interchangeably in the past, but they refer to related yet distinct properties of soils. To understand the nuance, we must define each term and examine their relationship.

1.2.1 Sodic soils: a chemical definition

A sodic soil is traditionally defined by a high proportion of sodium ions relative to other cations on the soil's exchange sites. Sodicity is quantified by measures such as the ESP (the percentage of the soil's cation exchange capacity occupied by sodium) or the SAR, calculated from soluble Na, Ca, and Mg in soil water. Historically, an ESP above 15% (or an SAR above ~13) was used to classify sodic soils (Richards 1954). However, different regions use different thresholds; in Australia, an ESP of ≥6% is often considered sodic for agronomic purposes (Northcote and Skene 1972). A sodic soil typically has low salinity (i.e. electrical conductivity of a saturated soil extract (EC_e) <4 dS/m) to distinguish it from saline-sodic soils. Importantly, sodicity is a chemical property and does not directly describe how soil behaves in water (Levy 2000; Rengasamy 2002).

1.2.2 Dispersive soils: a physical behaviour

A dispersive soil is defined by its behaviour when wetted. Clay particles detach and repel each other, causing the soil structure to collapse into a suspension. When

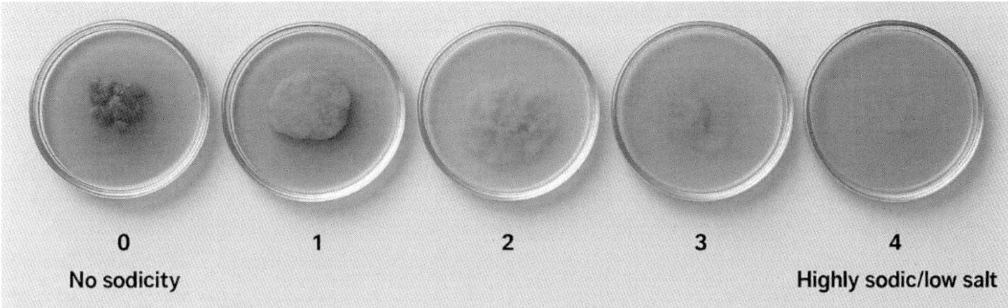

Fig. 1.1. Visual assessment of soil dispersion using the crumb test across a sodicity gradient (0–4). Petri dishes show the extent of clay dispersion when soil aggregates are immersed in deionised water. Grade 0 (no sodicity) shows intact aggregates with clear water, indicating stable structure. Grades 1–2 show increasing cloudiness, signifying moderate dispersion. Grades 3 and 4 exhibit extensive dispersion, with clay particles fully dispersed into solution, characteristic of highly sodic soils with low electrolyte concentration. This test provides a rapid field-based indicator of soil structural stability and sodicity severity.

placed in water, dispersive soil visibly clouds the water, indicating poor aggregate stability. This behaviour is typically tested using empirical methods such as the Emerson crumb test or the pinhole test (Emerson and Bakker 1973; Fig. 1.1). Dispersiveness is a physical trait with immediate implications: such soils are prone to erosion, crusting, and tunnel formation, even when sodium levels are moderate (Rengasamy *et al.* 2023).

1.2.3 Sodic soils versus dispersive soils

Although sodic soils and dispersive soils are related through the role of sodium, they are not synonymous categories. Many sodic soils are indeed dispersive, because excess exchangeable sodium typically promotes clay dispersion. However, it is crucial to recognise that not all sodic soils disperse, and not all dispersive soils are sodic. The overlap is significant but incomplete. Several factors account for this discrepancy:

- electrolyte concentration (salinity): the presence of dissolved salts in the soil water can counteract the dispersive effect of sodium. A soil with high ESP may not disperse if there is also a high concentration of calcium salts or other electrolytes in the soil solution, which help keep clay particles flocculated (clumped). For instance, field observations have documented soils with ESP as high as 25–30% that remain structurally stable because of elevated salinity or gypsum/calcium carbonate content (Quirk and Schofield 1955). In one case, a soil with an ESP of ~26% showed no dispersion due to stabilising factors like soluble salts and lime (calcium carbonate) in the profile (McIntyre 1979). Thus, a high sodium soil can escape dispersiveness if counterbalancing flocculating agents are present. Conversely, if a soil has very low salinity (e.g. under rainfall conditions where salts are leached out), even a modest amount of exchangeable sodium can induce dispersion (McNeal and Coleman 1966). The critical role of salt concentration is well documented: if the total electrolyte concentration falls below the critical flocculation concentration (CFC) of the system, clay particles will tend to disperse readily, even at lower ESP. This is why sodic soils in humid, leached environments often exhibit more dispersion problems than those in semi-arid, saline environments: the salt balance is as important as the sodium percentage (Dregne 1952)

- thresholds are not universal: there is no single universal ESP value at which a soil suddenly becomes dispersive. Research has shown a wide continuum of behaviour. Some soils begin to disperse at very low ESP (well below the old 15% threshold), especially under pure water conditions or agitation. Classic studies demonstrated that even at ESP 1–2%, clay dispersion and permeability loss can occur in distilled water (Shainberg *et al.* 1981). Field surveys have reinforced this point; for example, an analysis of 306 soil samples in south-eastern Australia found that ~17% of samples with an ESP <6% were dispersive (despite being classified as non-sodic by traditional criteria; McDonald *et al.* 2017; Tavakkoli *et al.* 2022). Conversely, the same studies noted that a few soils with an ESP >6% did not disperse. A larger dataset from northern New South Wales and Queensland similarly showed that 38% of soils with even mild sodicity (ESP ≥5%) were non-dispersive (McKenzie and Bryce 2023). These data underscore that ESP or SAR alone are imperfect predictors of dispersive behaviour. A soil's position on the sodicity spectrum does not guarantee a certain structural outcome. Consequently, one must be cautious in assuming 'sodic = dispersive' without further evidence (Fig. 1.2)
- cation type and soil mineralogy: the type of cations on the exchange complex influences dispersion. Sodium is highly dispersive, but other cations play a role too. Magnesium, for example, although a divalent cation like calcium, tends to act more like sodium in promoting dispersion (though to a lesser degree than sodium; Essington 2015). So a soil may have low ESP but high exchangeable magnesium, which can still result in poor structure. Soils with a high magnesium/sodium ratio (with relatively less calcium) often exhibit dispersion or at least weak structure, even if they do not meet the sodicity definition. Potassium, another monovalent cation, can also contribute to dispersion when present in high amounts relative to calcium. Recent work has expanded the sodicity concept to include these cationic effects: for example, the Cation Ratio of Soil Structural Stability (CROSS) incorporates sodium (Na^+), potassium (K^+), magnesium (Mg^{2+}), and calcium (Ca^{2+}) in a combined formula (Marchuk and Rengasamy 2011, 2012). Studies have found that such

composite indices correlate with dispersion better than SAR (which considers only Na relative to Ca+Mg). The implication is that a non-sodic soil based on Na percentage may still be dispersive if, say, K is abundant or Ca is low (Rengasamy *et al.* 2016; McDonald *et al.* 2020; Tavakkoli *et al.* 2022). Clay mineralogy also matters: soils dominated by swelling clays (smectites) are more prone to disperse under sodic conditions than those with kaolinite or iron oxides that cement particles together. Some clays are naturally stable (e.g. subplastic clays with high iron/aluminium content) and resist dispersion even at high ESP

- soil organic matter and other stabilisers: the presence of organic matter, soil biota (such as fungal hyphae), or carbonates can strengthen soil aggregates. Even a sodic soil may not disperse if it has plenty of organic binding agents or precipitated calcium carbonate (lime) that glues particles. Well-structured sodic soils with high organic matter can remain flocculated, whereas a low-organic sodic soil is much more vulnerable to dispersion. Many dispersive soils are observed to have low organic matter content, which exacerbates the dispersibility of their clays (Gupta *et al.* 1984; Barzegar *et al.* 2002).

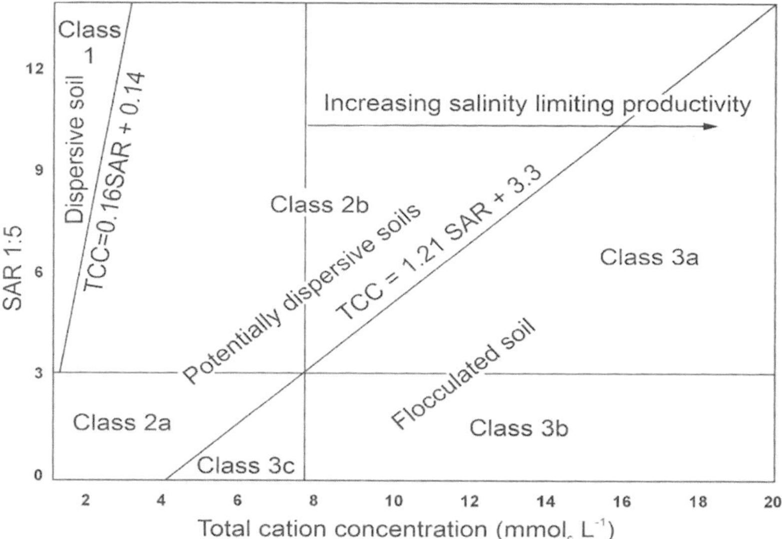

Fig. 1.2. Classification of soil dispersivity based on the sodium adsorption ratio of 1:5 extracts ($SAR_{1:5}$) and total cation concentration (TCC). The diagram delineates zones of soil behaviour using the relationship between $SAR_{1:5}$ and TCC, the latter representing the electrolyte concentration threshold above which clay dispersion does not occur. The empirical boundaries (i.e. TCC = 0.16 SAR + 0.14 and TCC = 1.21 SAR + 3.3) distinguish between dispersive (Class 1), potentially dispersive (Class 2a and 2b), and flocculated soils (Class 3a–c). Increasing salinity along the *x*-axis may suppress dispersion but also limit plant productivity. Adapted from Rengasamy *et al.* (1984).

In summary, 'sodic soil' refers to a soil with a high sodium content on its exchange complex, and this condition often predisposes the soil to dispersion problems. 'Dispersive soil' refers to a soil that actually exhibits clay dispersion *in situ* or in tests. A soil can be both sodic and dispersive, but it can also be sodic yet non-dispersive (if other factors mitigate dispersion) or non-sodic yet dispersive (if other factors induce dispersion). The distinction is more than semantic: it is practical. Sodicity is an inherent soil property, whereas dispersiveness is an outcome that affects soil behaviour and land management. Recognising this difference is the first step towards improved diagnosis of problem soils. The next section explains why this book advocates moving away from the term 'sodic soil' to the concept of 'dispersive soil' as the central focus, given the above complexities.

1.3 MOVING AWAY FROM 'SODIC SOIL' TO 'DISPERSIVE SOIL'

Given the nuanced relationship between sodicity and dispersion, soil scientists are increasingly calling for a shift in how we classify and manage sodic and dispersive soils. The traditional reliance on ESP/SAR thresholds to identify problematic sodic soils is now seen as too limited. Instead, there is a growing consensus that the key diagnostic criterion should be the soil's dispersive behaviour itself, not just its sodium level. This section elaborates on why a move from the term 'sodic soil' to 'dispersive soil' is more than just a change in vocabulary and instead represents an important change in perspective and practice.

1.3.1 Limitations of sodicity-based definitions

Decades of research have exposed fundamental problems with defining soil hazard purely by exchangeable sodium levels. Sumner (1993) famously argued that 'a single simple definition [of sodicity] is no longer possible' because soils once considered non-sodic can still exhibit the deleterious properties of sodic soils under the right conditions. One major limitation is the lack of a universal ESP threshold for structural problems, as discussed above. Soil behaviour varies continuously with electrolyte conditions and other factors, so any sharp cut-off (e.g. ESP >6%, >15%) is inherently arbitrary and context-dependent (Sumner and Naidu 1998). An ESP of 10% may be innocuous in a saline environment but dangerous in a leached soil with pure rainwater; the old definitions cannot capture this situational difference. Moreover, using ESP/SAR alone can be misleading: recent field data and laboratory studies (McKenzie and Bryce 2023) show that predicting dispersion from ESP is error-prone. That is, chemical indices cannot reliably substitute for a direct assessment of dispersion. The literature is replete with examples where management based purely on sodicity measures failed to solve the soil's structural issues, or where soils deemed 'safe' due to moderate ESP unexpectedly collapsed after rainfall. As a result, relying on ESP/SAR as proxies for dispersion is no longer recommended by experts. The push is to test directly for dispersion or to use improved indices that better capture dispersion risk (McIntyre 1979).

1.3.2 Rationale for a dispersive soil focus

In practical terms, what ultimately matters to land managers, engineers, and ecologists is whether a soil will retain its structure when wet or not. A 'dispersive soil' focus zeroes in on this outcome. By identifying dispersive soils (via behavioural tests or observations) we directly pinpoint the soils that are prone to erosion, gullies, tunnelling, and poor crop performance, regardless of why they disperse. This approach has several advantages:

- It captures all causes of dispersion, not just sodium. For instance, if a soil disperses due to unusual ionic composition (say high K or Mg) or low organic matter, a dispersive-soil approach would still flag it, whereas a sodicity criterion may miss it. Thus, a dispersive-soil approach is more holistic in diagnosing problem soils.
- It avoids false positives/negatives associated with sodicity metrics. As discussed, not all high-ESP soils are problematic, and not all low-ESP soils are benign. Focusing on dispersion sidesteps this issue: the soil either disperses in a standard test or it does not. If it does, it is treated as a concern, prompting investigation of causes (which may include high ESP, but also other factors). If it does not disperse, one may avoid unnecessary remediation even if ESP is high (e.g. saving resources by not applying amendments to a stable sodic soil that is not actually causing structural issues). This nuance is important for policymakers who must target interventions cost-effectively.
- It links directly to end-user observations. Farmers and engineers often first notice symptoms like muddy water, poor infiltration, or collapsing earth structures. By framing the issue in terms of dispersive soils, we align scientific criteria with these observable phenomena. In fact, simple field tests for dispersion (like the crumb test, where a soil fragment in water is observed for milkiness) can empower practitioners to identify dispersive soils on-site without waiting for laboratory sodium measurements. Such tests are now recommended as part of standard soil surveys in some regions, reflecting the shift towards behaviour-based identification.
- It facilitates better management strategies. When a soil is recognised as dispersive, management can be tailored to address why it disperses (e.g. leaching salts, adding gypsum to displace exchangeable sodium, incorporating organic

matter to promote aggregation, or using polyacrylamide (PAM) to stabilise soil structure). These strategies are detailed later in this book. The key point is that by diagnosing the actual problem (dispersion), one can more directly choose an appropriate solution. In contrast, managing 'sodicity' alone may lead to generic treatments that do not fully resolve the issue if other factors are at play. For instance, adding gypsum to a sodic soil will add electrolytes and calcium, which usually helps flocculate clays, but if the soil's dispersion was partly due to very low organic matter or a transient salinity drop, one may need to also address those aspects (like cover cropping for organic matter or irrigation management for salinity). A dispersive-soil framework encourages this comprehensive view.

1.3.3 Evidence of paradigm shift

The academic and professional soil community has increasingly embraced the change to a dispersive soil focus in recent years. Revised soil classification schemes and manuals now emphasise dispersion characteristics alongside or in place of traditional sodicity classes. In Australia, for example, field manuals for soil management explicitly differentiate 'dispersive (sodic) soils' and warn that ESP is not a reliable stand-alone indicator (McKenzie and Bryce 2023). The field manuals advocate using direct dispersion tests (e.g. the Emerson aggregate test or laboratory aggregate stability in water tests) to identify dispersive layers in soil profiles. Researchers have developed indices like the Electrochemical Stability Index (ESI), which combines ESP with electrical conductivity in a single value, to better predict dispersion potential (Marchuk and Rengasamy 2012). Others have proposed the aforementioned CROSS index to include potassium's effect, highlighting that soil science is moving beyond the old ESP/SAR paradigm. Even the language in the scientific literature is shifting: instead of focusing only on 'sodic soils', as was common in mid-20th century literature, recent papers and reviews talk about 'dispersive soils' and their management (Vakili *et al.* 2024). This reflects a more nuanced understanding that it is the dispersive behaviour (and its consequences) that we ultimately care about.

In light of these developments, this book adopts the term 'dispersive soils' as a central concept. We will, of course, discuss sodicity (because sodium remains a primary driver of dispersion) but always with the recognition that sodium levels are part of a bigger picture of soil behaviour. By moving away from a narrow focus on 'sodic soils' towards the broader concept of dispersive soils, we aim to equip the reader with a more powerful diagnostic lens. This shift enables a clearer interpretation of soil test data and more effective decision-making in land management.

In the chapters that follow, we begin with Chapter 2, which delves into the mechanisms of clay dispersion and the fundamental physicochemical processes underpinning this behaviour. Chapter 3 discusses how dispersive soils are diagnosed and tested, including field-based and laboratory techniques. Chapter 4 focuses on the agricultural consequences of dispersive soils and outlines current and emerging management practices. Chapter 5 addresses the interaction between irrigation practices and dispersive soils, especially how water quality and management strategies influence soil stability. Chapter 6 explores the broader impacts of dispersive soils on ecosystems, environments, and infrastructure, and presents potential mitigation strategies. Chapter 7 provides detailed guidance on how dispersive soils are identified and classified across different landscapes and soil types. Finally, Chapter 8 discusses key knowledge gaps, future research priorities, and interdisciplinary opportunities to advance our understanding and management of dispersive soils in a changing world.

Overall, the transition from the concept of 'sodic soil' to that of 'dispersive soil' represents a maturation in soil science: from a static, chemistry-based definition to a dynamic, behaviour-based understanding. This introductory chapter has set the stage by explaining why this shift is necessary and how it benefits both scientific inquiry and practical problem solving. With this perspective in mind, we are prepared to advance into a deeper exploration of dispersive soils in the coming chapters, always keeping in focus the ultimate goal: improving land use outcomes by addressing the real causes of soil degradation.

1.4 REFERENCES

Barzegar AR, Yousefi A, Daryashenas A (2002) The effect of addition of different amounts and types of organic materials on soil physical properties and yield of wheat. *Plant and Soil* **247**, 295–301. doi:10.1023/A:1021561628045

Dregne HE (1952) Alkali soils, their formation, properties, and reclamation. *Soil Science Society of America Journal* **16**, 372. doi:10.2136/sssaj1952.03615995001600040015x

Emerson WW, Bakker AC (1973) The comparative effects of exchangeable calcium, magnesium, and sodium on some physical properties of red-brown earth subsoils. II. The spontaneous dispersion of aggregates in water. *Soil Research* **11**, 151–157. doi:10.1071/SR9730151

Essington ME (2015) 'Soil and water chemistry: an integrative approach.' (CRC Press: Boca Raton, FL)

Gupta RK, Bhumbla DK, Abrol IP (1984) Effects of sodicity, pH, organic matter and calcium carbonate on the dispersion behaviour of soils. *Soil Science* **137**, 245–251.

Levy GJ (2000) Sodicity. In 'Handbook of soil science.' (Ed. ME Sumner) pp. 27–55. (CRC Press: Boca Raton, FL)

Marchuk A, Rengasamy P (2011) Clay behaviour in suspension is related to the ionicity of clay–cation bonds. *Applied Clay Science* **53**, 754–759. doi:10.1016/j.clay.2011.05.019

Marchuk A, Rengasamy P (2012) Threshold electrolyte concentration and dispersive potential in relation to CROSS in dispersive soils. *Soil Research* **50**, 473–481. doi:10.1071/SR12135

McDonald GK, Tavakkoli E, Cozzolino D, Banas K, Derrien M, Rengasamy P (2017) A survey of total and dissolved organic carbon in alkaline soils of southern Australia. *Soil Research* **55**, 617–629. doi:10.1071/SR16237

McDonald GK, Tavakkoli E, Rengasamy P (2020) Commentary: bread wheat with high salinity and sodicity tolerance. *Frontiers in Plant Science* **11**, 1194. doi:10.3389/fpls.2020.01194

McIntyre DS (1979) Exchangeable sodium, subplasticity and hydraulic conductivity of someAustralian soils. *Australian Journal of Soil Research* **17**, 115–120.

McKenzie D, Bryce A (2023) Dispersive soil manual. Managing dispersive soils: practicalities and economics – Northern region. Grains Research and Development Corporation. Available at https://grdc.com.au/resources-and-publications/all-publications/publications/2023/dispersive-soil-manual [verified 14 July 2025].

McNeal BL, Coleman NT (1966) Effect of solution composition on soil hydraulic conductivity. *Soil Science Society of America Journal* **30**, 308–312. doi:10.2136/sssaj1966.03615995003000030007x

Naidu R, Sumner ME, Rengasamy P (1995) 'Australian sodic soils: distribution, properties, and management.' (CSIRO Publications: Melbourne, Vic.)

Northcote KH, Skene JKM (1972) 'Australian soils with saline and sodic properties.' (CSIRO: Melbourne, Vic.)

Quirk JP, Schofield RK (1955) The effect of electrolyte concentration on soil permeability. *Journal of Soil Science* **6**, 163–178. doi:10.1111/j.1365-2389.1955.tb00841.x

Rengasamy P (2002) Transient salinity and subsoil constraints to dryland farming in Australian sodic soils: an overview. *Australian Journal of Experimental Agriculture* **42**, 351–361. doi:10.1071/EA01111

Rengasamy P, Marchuk A (2011) Cation ratio of soil structural stability (CROSS). *Australian Journal of Soil Research* **49**, 280–285. doi:10.1071/SR10105

Rengasamy P, Green RSB, Ford GW, Menhanni AH (1984) Identification of dispersive behaviour and the management of red-brown earths. *Australian Journal of Soil Research* **22**, 413–431.

Rengasamy P, Tavakkoli E, McDonald GK (2016) Exchangeable cations and clay dispersion: net dispersive charge, a new concept for dispersive soil. *European Journal of Soil Science* **67**, 659–665. doi:10.1111/ejss.12369

Rengasamy P, McDonald GK, Tavakkoli E (2023) Managing multiple constraints to crop production in dispersive soils. In 'Soil constraints and productivity.' (Eds N Bolan, MB Kirkham) pp. 319–341. (CRC Press: Boca Raton, FL)

Richards LA (1954) Diagnosis and improvement of saline and alkali soils. *Soil Science* **78**, 154. doi:10.1097/00010694-195408000-00012

Shainberg I, Rhoades JD, Prather RJ (1981) Effect of low electrolyte concentration on clay dispersion and hydraulic conductivity of a sodic soil. *Soil Science Society of America Journal* **45**, 273–277. doi:10.2136/sssaj1981.03615995004500020009x

Sumner ME (1993) Sodic soils – new perspectives. *Australian Journal of Soil Research* **31**, 683–750. doi:10.1071/SR9930683

Sumner ME, Naidu R (1998) 'Sodic soils: distribution, properties, management, and environmental consequences.' (Oxford University Press: New York, NY)

Tavakkoli E, Uddin S, Rengasamy P, McDonald GK (2022) Field applications of gypsum reduce pH and improve soil C in highly alkaline soils in southern Australia's dryland cropping region. *Soil Use and Management* **38**, 466–477. doi:10.1111/sum.12756

Vakili AH, Salimi M, Keskin İ, Jamalimoghadam M (2024) A systematic review of strategies for identifying and stabilizing dispersive clay soils for sustainable infrastructure. *Soil and Tillage Research* **239**, 106036. doi:10.1016/j.still.2024.106036

2

Formation and distribution of dispersive soils

2.1 INTRODUCTION

Dispersive soils, characterised by the spontaneous breakdown of soil aggregates into individual clay particles upon wetting, represent a major constraint to sustainable land use across agricultural and environmental systems. First formally recognised by Puri and Keen (1925), the behaviour of dispersive soils has challenged soil scientists for nearly a century. Historically, dispersive soils were broadly classified under categories such as 'sodic soils', 'Solonetz', or 'alkali soils' (Kelley 1951; Szabolcs 1989), with emphasis placed primarily on sodium (Na^+) as the dominant destabilising cation.

Contemporary research has expanded this view by demonstrating that other cations, including potassium (K^+), magnesium (Mg^{2+}), and ammonium (NH_4^+), can also promote clay dispersion under specific conditions (Marchuk and Rengasamy 2012). This shift underscores the need to consider multiple interacting factors (e.g. clay mineralogy, soil pH, organic matter content, and electrolyte concentration) when assessing dispersive behaviour. As explored in Chapter 3, the concept of 'net dispersive charge' provides a more comprehensive framework for diagnosing dispersion risk, moving beyond conventional reliance on exchangeable sodium percentage (ESP) or sodium adsorption ratio (SAR) alone.

This chapter outlines the key processes involved in the formation of dispersive soils, beginning with an examination of natural and anthropogenic salinisation processes and pathways. These include groundwater-associated salinity, non-groundwater-associated (transient) salinity, and irrigation-induced salinity, each of which contributes uniquely to soil structural degradation. This chapter also introduces the physicochemical basis for clay dispersion, differentiates between dispersive, saline–dispersive, and saline soils, and provides an overview of the global distribution and mapping limitations of dispersive soils, with particular emphasis on Australian landscapes.

2.2 PROCESSES INVOLVED IN THE FORMATION OF DISPERSIVE SOILS

The genesis of dispersive soils is closely linked to salinisation, namely the accumulation of soluble salts in the soil profile to concentrations that adversely affect soil structure, hydraulic properties, and biological function. Salinisation may occur via natural (primary) or anthropogenic (secondary) pathways, and its effects are compounded by shifts in vegetation, climate, and land use.

As salts accumulate, they alter both the ionic strength of the soil solution and the composition of the cation exchange complex. These changes can weaken interparticle bonding, reduce flocculation, and promote clay dispersion. The resulting structural degradation impairs porosity, water infiltration, and root penetration while increasing erosion and surface crusting. Globally, salinity affects approximately one billion hectares (~7% of the Earth's land surface) and impacts nearly one-third of all irrigated land (Hopmans *et al.* 2022). In dryland regions, particularly in Australia, limited leaching and rising water tables have accelerated the spread of both salinity and dispersive soil conditions (Rengasamy 2006).

2.2.1 Soil salinisation pathways

Salinisation is broadly categorised into primary and secondary processes based on origin and timescale, as detailed below.

- Primary salinity arises from long-term geological and atmospheric processes, such as marine transgressions, weathering of parent materials, and deposition of marine aerosols or aeolian dust. In low-relief, poorly drained landscapes, salts accumulate in the soil profile over centuries due to limited leaching. Coastal environments may also experience salinity through saline groundwater discharge or seawater intrusion.
- Secondary salinity results from human-induced modifications to the hydrological cycle. Irrigation using saline or marginal-quality water, excessive application of fertilisers and industrial effluents, and inadequate drainage are major contributors. Land clearing and the replacement of deep-rooted perennial vegetation with shallow-rooted annual crops reduce evapotranspiration and increase recharge, mobilising salt stores previously stabilised at depth. The redistribution of these salts can elevate soil sodicity and create conditions conducive to dispersion, especially in soils with low permeability.

Both salinisation pathways create vertical and lateral gradients in salt concentration and ionic composition that influence aggregate stability and dispersion potential.

2.2.2 Groundwater-associated salinity

Groundwater-associated salinity, commonly referred to as 'seepage salinity', is typically observed in landscape discharge zones, where rising water tables transport dissolved salts into the root zone and surface soil layers. When the water table rises to within ~1.5 m of the surface, capillary rise becomes a dominant process, drawing water and solutes upward. Subsequent evaporation at or near the surface concentrates salts in the upper profile. Prior to extensive agricultural development, deep-rooted perennial vegetation maintained stable water tables through high transpiration rates. However, the conversion of these ecosystems to shallow-rooted cropping systems has disrupted the water balance. Increased recharge raises groundwater levels, leading to salt accumulation in surface soils and further driving dispersion, particularly where sodium dominates the cation exchange complex (Rengasamy 2006).

This process often results in soil crusting, reduced infiltration, waterlogging, and widespread structural degradation. Lateral groundwater seepage along slopes and valley footslopes exacerbates salinity spread and promotes the formation of saline–dispersive mosaics across the landscape.

2.2.3 Non-groundwater-associated (transient) salinity

Transient salinity occurs in deep water table landscapes, where salts accumulate in the subsoil due to low rainfall and insufficient leaching. This form of salinity is not driven by rising groundwater but by the long-term deposition of soluble salts from atmospheric sources and mineral weathering. In many Australian regions, 'salt bulges' have been identified at depths ofbetween 4 and 10 m, where salts accumulate over time in slowly draining horizons (Fig. 2.1; Holmes 1960; Lawrie 2005).

During episodic rainfall events, these salts may be mobilised upward into the root zone, causing fluctuating osmotic stress, ion toxicity, and, in sodium-dominated systems, dispersive behaviour. This dynamic salinity pattern is particularly damaging in dryland cropping areas, where root-zone water and nutrient availability are already constrained (Fig. 2.2).

It is estimated that ~67% of Australian agricultural land is affected by transient salinity to some degree (Rengasamy 2002; Barrett-Lennard *et al.* 2016). In addition to sodium, transient salinity is often associated with elevated concentrations of boron, bicarbonate, aluminium, manganese, and iron, which can further restrict plant growth. Climate projections indicating decreased rainfall and higher evapotranspiration suggest a likely increase in the severity and frequency of transient salinity events in semi-arid regions. Conversely, in regions experiencing higher rainfall, elevated recharge may drive transitions towards groundwater-induced salinity.

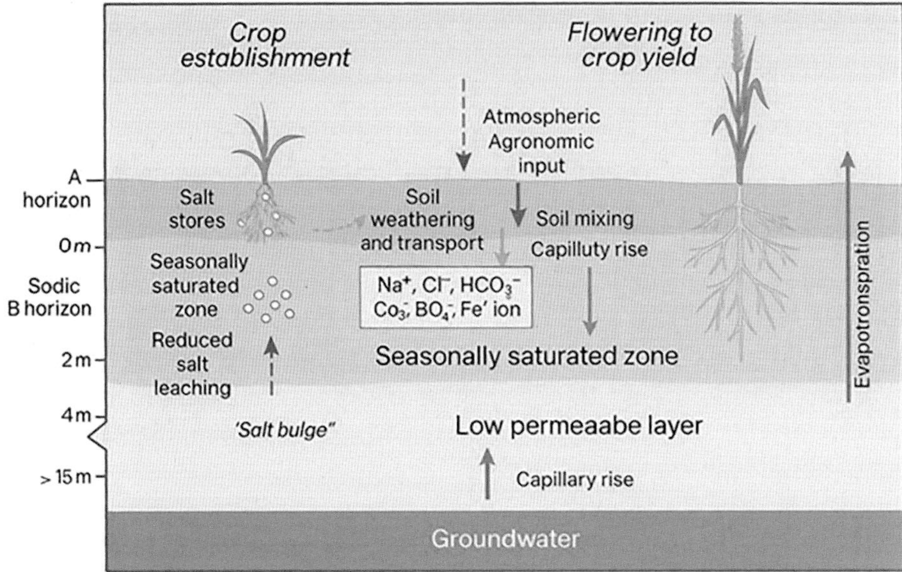

Fig. 2.1. Schematic representation of transient salinity dynamics in dispersive soils. The diagram illustrates key soil processes influencing salinity and plant growth from early crop establishment to reproductive stages. During seasonal saturation, salts accumulate in the subsoil due to limited leaching and upward capillary rise, particularly above low-permeability sodic horizons. Atmospheric and agronomic inputs, combined with soil weathering, contribute to the accumulation of soluble ions such as Na^+, Cl^-, HCO_3^-, CO_3^{2-}, BO_3^-, and Fe^{2+}/Fe^{3+}. Capillary rise transports these ions towards the root zone, where evapotranspiration concentrates them further. Salt bulges form above low-permeability layers, exacerbating dispersion and root-zone constraints. Adapted from Rengasamy (2002).

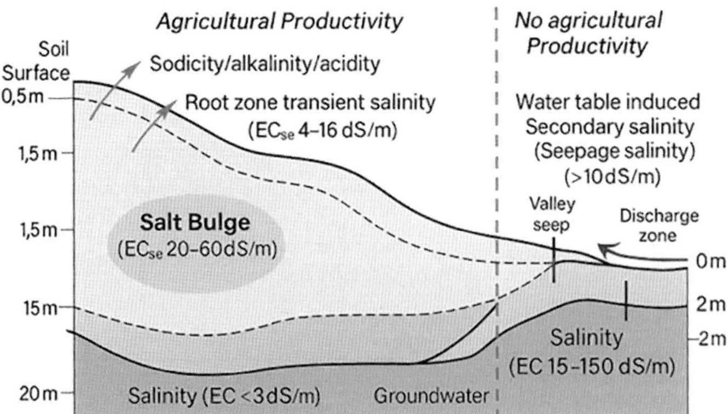

Fig. 2.2. Conceptual illustration of dryland salinity types in Australian landscapes, showing the vertical and lateral distribution of salinity processes. Primary salinity originates from underlying saline groundwater, whereas transient salinity in the root zone and water table-induced secondary salinity are influenced by landscape position and hydrological fluxes. The salt bulge and discharge zones highlight critical areas of salt accumulation and potential land degradation. Adapted from Rengasamy (2002).

2.2.4 Irrigation-induced salinity

Irrigation-induced salinity is now a leading cause of salt accumulation and structural degradation in intensively managed agricultural systems. Salinity arises when irrigation water, drawn from surface sources, bores, or recycled wastewater, contains elevated salt levels and is applied without sufficient leaching. In clay-rich dispersive soils, limited hydraulic conductivity and high evaporative demand exacerbate salt concentration near the surface.

As early as the 1990s, the Food and Agriculture Organization of the United Nations (FAO) reported that 20–30 million hectares of irrigated land was severely salinised, with an additional 80 million hectares moderately affected (FAO 1991). The scale of this problem has since grown, particularly in semi-arid regions where irrigation is essential for crop production. In Australia, the practice of pumping saline groundwater for reuse in irrigation has, in many cases, worsened surface salinity and sodicity (Rengasamy and Olsson 1993). Similarly, the use of recycled water and industrial effluents, often high in sodium, bicarbonate, and residual alkalinity, poses significant structural risks.

The severity of irrigation-induced dispersion is influenced by water quality parameters such as electrical conductivity (EC), SAR, and residual sodium carbonate. Poor-quality water can displace calcium and magnesium from clay surfaces, reducing aggregate stability. The method of water application also plays a role; for instance, flood irrigation tends to exacerbate salt stratification compared with more targeted systems such as drip irrigation.

Numerous studies (Shainberg and Shalhevet 1984; Rengasamy and Olsson 1993; Mohanavelu *et al.* 2021) have documented how irrigation water quality affects infiltration, structure, and plant-available water. As climate change and water scarcity increase reliance on marginal water sources, the risk of both salinisation and dispersion intensifies. Addressing these issues requires integrated approaches that consider soil–water–plant interactions, drainage design, and irrigation management to maintain long-term soil function and crop productivity.

2.3 FORMATION MECHANISMS OF DISPERSIVE SOILS

The formation of dispersive soils is fundamentally governed by electrochemical interactions between the soil solid phase, comprising clay minerals and adsorbed cations, and the surrounding soil solution. These interactions control the stability of soil aggregates by modulating the balance between interparticle attractive and repulsive forces. When repulsive forces exceed attractive forces, clay particles disperse from aggregates, resulting in structural degradation. This phenomenon is central to the behaviour of dispersive soils and underpins their management challenges.

The concept of dispersive charge, as used in this chapter, refers to the net potential for clay dispersion arising from the specific composition and electrochemical behaviour of exchangeable cations adsorbed onto clay surfaces. Unlike cation exchange capacity, which quantifies the total exchangeable cations, dispersive charge reflects the qualitative effect of these cations on colloidal stability. Monovalent cations such as Na^+ and K^+, due to their low charge density and large hydrated radii, create strong repulsive forces between clay particles. These ions reduce interparticle bridging and promote dispersion. In contrast, divalent cations such as calcium (Ca^{2+}) and Mg^{2+} have a flocculating effect by bridging adjacent clay particles, enhancing structural stability.

Balancing this is the flocculating charge, which derives from the ionic strength and cation composition of the soil solution. Soluble cations in the soil water, particularly divalent species, can shield the negative surface charges of clays, reducing electrostatic repulsion and promoting flocculation. This shielding effect, a key principle in DLVO (Derjaguin–Landau–Verwey–Overbeek) theory, is sensitive to both the concentration and valency of ions in solution. When the flocculating charge is sufficiently high, dispersion is suppressed. Conversely, if the dispersive charge from the exchange complex exceeds the flocculating capacity of the soil solution, the net dispersive force becomes positive, triggering clay detachment and aggregate breakdown.

The evolution of dispersive behaviour is closely linked to the accumulation and composition of salts within the soil profile. Salts derived from weathering, irrigation, or rising saline groundwater dissolve into the soil solution and contribute cations such as Na^+, Ca^{2+}, Mg^{2+}, and K^+. These cations compete for adsorption sites on negatively charged clay surfaces. The dominance of sodium in the exchange complex, a hallmark of sodic soils, amplifies the dispersive charge. Meanwhile, cations that remain in the soil solution determine the strength of the flocculating charge. Thus, dispersive soils often emerge in environments with both sodic and saline features, where the specific balance of cation types across solid and liquid phases dictates soil structure outcomes.

It is important to note that dispersive, saline–dispersive, and saline soils represent a continuum, not discrete categories. Dispersive soils are typically characterised by low electrolyte concentrations and high sodium saturation on the exchange complex. Saline–dispersive soils contain enough soluble salts to partially suppress dispersion under certain conditions, but remain structurally vulnerable due to their sodic nature. In contrast, non-dispersive saline soils maintain structural integrity because their high electrolyte concentrations exceed the dispersive threshold. However, leaching of salts without concurrent adjustment of exchangeable cation composition can destabilise these soils, leading to a transition into dispersive states.

Soil pH further modulates dispersion potential. In acidic soils (pH <6), high concentrations of exchangeable aluminium (Al^{3+}) and hydrogen (H^+) introduce unique electrochemical behaviours that can influence clay stability, although dispersion is generally less pronounced. Soils with near-neutral pH (6–8) exhibit more stable aggregation, particularly when calcium dominates the exchange complex. However, as pH exceeds 8, particularly in sodic soils, dispersion risk increases sharply. High pH enhances sodium activity and promotes the formation of carbonate and bicarbonate ions, which can reduce soluble calcium through precipitation and further destabilise the soil structure. In highly alkaline conditions (pH >9), this process is often accompanied by swelling, colloidal mobility, and surface crusting.

The interplay among cation composition, electrolyte concentration, pH, and ionic interactions defines the mechanistic basis for dispersive soil behaviour. Understanding this balance is essential for diagnosing dispersion risk and designing management strategies. For instance, remediation approaches such as gypsum application target this equilibrium by increasing the supply of flocculating divalent cations (e.g. Ca^{2+}), thereby shifting soils towards a stable, aggregated state.

In summary, the formation of dispersive soils is not a singular event but a dynamic process driven by the relative dominance of dispersive and flocculating forces, the history of salt accumulation and leaching, and the evolving chemistry of the soil environment. The following chapters explore diagnostic techniques and targeted amelioration strategies based on this mechanistic understanding.

2.4 CHARACTERISTICS OF DISPERSIVE SOILS

Dispersive soils are defined by their structural instability, particularly the tendency of their clay particles to disperse into suspension when wet. This behaviour is fundamentally governed by a positive net dispersive charge (i.e. when the dispersive charge from exchangeable cations exceeds the flocculating charge provided by ions in the soil solution; Rengasamy *et al.* 2016). Although the specific exchangeable cation composition may vary, the key threshold lies in the electrochemical imbalance that favours particle repulsion over cohesion. These soils typically have low soluble salt concentrations, and consequently the flocculating capacity of the soil solution is insufficient to stabilise aggregates, resulting in the breakdown of soil structure.

Dispersive soils are commonly characterised by very low electrolyte concentrations and a dominance of monovalent cations such as Na^+ on the exchange complex. As a result, the osmotic effects commonly associated with salinity are often minimal, yet the structural degradation is profound.

The consequences of such degradation are multifaceted. Water infiltration and retention are significantly reduced, aeration is compromised, and thermal and mechanical properties are altered. These changes adversely affect seedling emergence, root elongation, and water and nutrient uptake. As the soil dries, dispersed clays can form dense, hard-setting layers with high bulk density. Surface crusting further impedes germination, while subsoil compaction restricts root penetration and limits access to deeper moisture reserves. The availability and utilisation of 'green water' (i.e. water stored in the soil profile and available to crops) are markedly diminished, reducing yield potential in dryland cropping systems.

Beyond effects on crop performance, the physical instability of dispersive soils has broader landscape and engineering implications. The detachment and transport of clay particles during rainfall events increase the risk of sheet, rill, and especially gully erosion. In many cases, tunnel erosion, subsurface piping caused by dispersive clays, is a major concern. These processes accelerate land degradation and sedimentation in waterways. Furthermore, the low bearing strength and high swell–shrink potential of these soils can undermine the geotechnical stability of infrastructure, leading to subsidence, cracking of foundations, and landslides, particularly on slopes or under fluctuating moisture conditions. Thus, the relevance of dispersive soils extends well beyond agricultural productivity into environmental sustainability and civil engineering.

2.4.1 Saline–dispersive soils

In many landscapes, dispersive soils may also contain appreciable concentrations of soluble salts, resulting in what are referred to as saline–dispersive soils. These soils maintain a positive net dispersive charge but have elevated levels of electrolytes in the soil solution. Under such conditions, the salts may provide temporary flocculating effects, partially suppressing dispersion. However, if the dispersive charge remains greater than the flocculating charge, clay dispersion still occurs, although the extent may vary seasonally depending on moisture dynamics and salt concentration. This dynamic behaviour gives rise to the phenomenon of transient salinity, as first described by Rengasamy (2002), in which leaching, rainfall, and evapotranspiration cycles alter the balance of salts and the structural condition of the soil.

Saline–dispersive soils exhibit a dual set of challenges: structural instability from clay dispersion and osmotic stress or ion-specific toxicity from accumulated salts. In addition to impaired infiltration and compaction, plant growth may also be constrained by salt-induced physiological effects, including reduced water uptake and nutrient imbalances. In some cases, toxic concentrations of sodium, boron, bicarbonate, or trace metals such as manganese and iron may exacerbate plant stress, particularly in sensitive species or under drought conditions. As rainfall dilutes or redistributes salts within the profile, dispersion risk may increase, especially if leaching lowers the electrolyte concentration without altering the exchangeable cation balance. Thus, saline–dispersive soils represent a transitional state with highly variable behaviour influenced by both seasonal and management factors.

2.4.2 Non-dispersive saline soils

In contrast to the previously described soil categories, non-dispersive saline soils contain sufficient soluble salts to ensure that the flocculating charge outweighs the dispersive charge. As a result, soil aggregates remain intact despite the presence of dispersive cations on the exchange complex. These soils are structurally stable under current conditions, although they may impose physiological constraints on plant growth through osmotic effects and ion toxicity. Sodium chloride is often the dominant salt in such environments, but other compounds, such as sulfates, bicarbonates, and borates, may also be present.

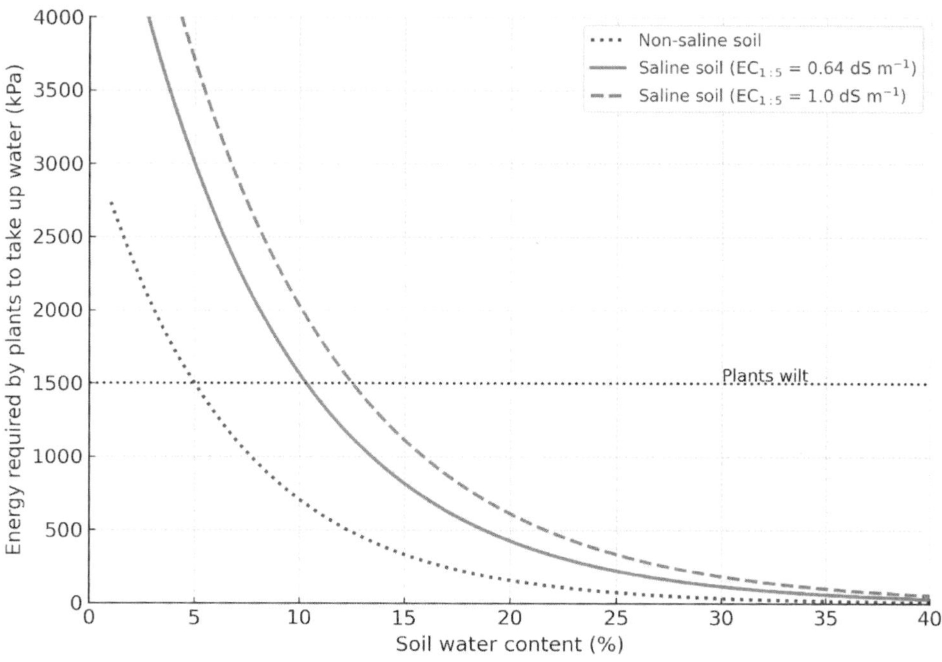

Fig. 2.3. Energy (kPa) required by plants to take up water as a function of soil water content and salinity, expressed by electrical conductivity of 1:5 extract ($EC_{1:5}$). Increasing salinity levels ($EC_{1:5}$ = 0.64 and 1.0 dS m^{-1}) substantially raise the energy demand for plant water uptake at any given soil moisture level, thereby reducing plant-available water. The threshold for permanent wilting is indicated at 1,500 kPa. Adapted from Rengasamy (2006).

A critical consideration in evaluating non-dispersive saline soils is that laboratory measurements of salinity, commonly conducted under saturated conditions, often underestimate in-field salt concentrations. Under natural field conditions, soil moisture is usually closer to field capacity, and evaporative losses can further concentrate salts in the root zone. This leads to elevated osmotic pressures that reduce the availability of water to plants. Plant water uptake is influenced by both matric and osmotic potentials, and the combined water potential determines the effective moisture availability. The soil's texture and mineralogy modulate these effects; for example, clay-rich soils may retain more water overall, but a substantial portion of the water can be held at tensions beyond the capacity of plant roots to extract. These interactions are explored in greater depth by Rengasamy (2010) and illustrated in Fig. 2.3.

It is important to emphasise that the structural stability of non-dispersive saline soils is not permanent. Leaching of salts due to rainfall or irrigation can reduce the flocculating charge below the critical threshold needed to suppress

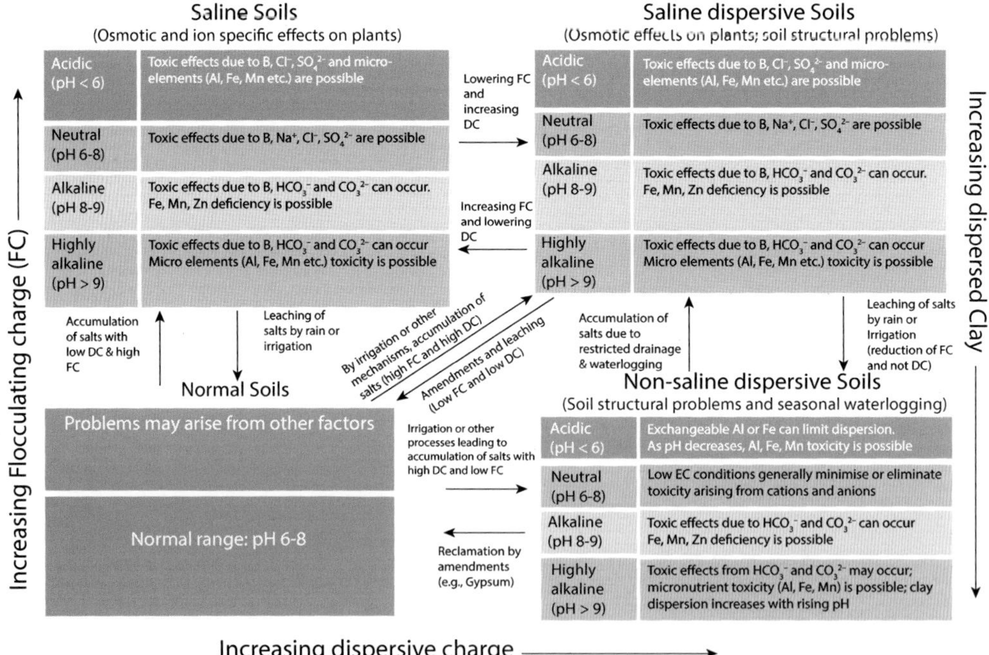

Fig. 2.4. Categories and dynamics of dispersive soils as related to dispersive charge, flocculating charge, and pH. The schematic illustrates four major soil types, namely normal, saline, saline–dispersive, and non-saline dispersive soils, classified based on their relative dispersive and flocculating charge. The diagram highlights the influence of pH on ion toxicity, clay dispersion, and soil structural stability, showing how cation composition, salt leaching, and chemical amendments can shift soils across categories. Arrows indicate transitional processes such as leaching, accumulation, and pH adjustment, with implications for soil management and reclamation.

dispersion. When this occurs, soils, particularly those with sodium- or potassium-dominated exchange complexes, can rapidly transition into a dispersive state. Moreover, soil pH further influences these dynamics by modulating cation activity and solubility. Under highly alkaline conditions, increased sodium activity and elevated concentrations of carbonate and bicarbonate ions exacerbate dispersion risks and contribute to nutrient imbalances. These interrelationships are conceptually summarised in Fig. 2.4 and are examined in detail in Chapter 3.

2.5 GLOBAL DISTRIBUTION OF DISPERSIVE SOILS

A fundamental limitation in our current understanding of the global extent of dispersive soils is the reliance on surrogate indicators, primarily exchangeable sodium percentage (ESP) and electrical conductivity (EC), rather than direct assessments of clay dispersion. Although numerous scientists have long advocated for the use of dispersion tests as the primary diagnostic criterion for sodic and dispersive soils (e.g. Puri and Keen 1925; Emerson 1967; North 1976; Rengasamy 2002), classification and mapping efforts continue to depend largely on indirect proxies. Although these metrics are useful in identifying salt-affected soils, they often fail to capture the full extent and variability of dispersive behaviour across landscapes.

Direct measurement of clay dispersion, or, more recently, quantification of net dispersive charge and flocculating capacity (as discussed in Chapter 3), provides a much more accurate picture of soil structural vulnerability. However, in the absence of globally harmonised methods and datasets, we must rely on the available information on saline and sodic soils to infer the probable distribution of dispersive soils. This approach remains imperfect, because it does not distinguish between sodic soils that are structurally stable and those that are prone to severe dispersion. Nonetheless, it provides a first-order approximation of the geographical and pedological zones where dispersive soils are most likely to occur.

The challenge is not a new one. As noted in previous sections, soil salinisation, closely linked to sodicity and dispersion, has shaped the trajectory of human civilisation for millennia. Historical records of land abandonment in southern Mesopotamia and the Tigris–Euphrates valley highlight the scale at which salt accumulation and related structural degradation can lead to social collapse (Jacobsen and Adams 1958; Hillel 2005). In the modern era, salt-affected soils are found in more than 100 countries and span every inhabited continent. Although much attention has been focused on arid and semi-arid climates, where salinity and sodicity are most severe, dryland salinity is increasingly emerging as a management concern even in temperate and humid regions due to changes in land use and hydrological balances.

All soils contain some soluble salts; however, when climatic and edaphic conditions favour the concentration of these salts beyond thresholds of agronomic or environmental concern, salinisation becomes a form of land degradation. The assumption that salinity is confined to arid environments is no longer valid. For instance, in the Canadian Prairies, ~4.5 million hectares of dryland cropping systems are affected by salinity (Wiebe *et al.* 2005). Salts responsible for these effects include not only sodium chloride, but also other compounds such as sulfates, carbonates, bicarbonates, borates, and various cations, including Ca^{2+}, Mg^{2+}, K^{+}, and Fe^{2+}/Fe^{3+} (Szabolcs 1989).

Szabolcs (1989) provided one of the earliest global overviews of salt-affected soils using FAO/UNESCO maps and national inventories. According to these data, more than 424 million hectares of surface soils (0–30 cm depth) and ~833 million hectares of subsoils (30–100 cm depth) are affected by salinity. These soils are found across all continents, with more than two-thirds concentrated in arid and semi-arid zones. Notably, approximately one-third of the world's irrigated land, equivalent to around 70 million hectares, is salt-affected (FAO and ITPS 2015). Unfortunately, global estimates of sodic and dispersive soils remain based on dated inventories, and new high-resolution assessments using modern soil behaviour indicators are urgently needed (Hopmans *et al.* 2022).

2.5.1 Classification of sodic soils in different countries

Global efforts to classify and map sodic soils are complicated by inconsistent national thresholds, terminology, and survey methods. Definitions of sodicity vary significantly across countries, impeding cross-regional comparisons and the development of harmonised global maps of sodic and dispersive soils. In the US, sodic soils are defined as those with a natric horizon and an ESP >15% (Soil Survey Staff 1992). In contrast, Australian classifications apply lower thresholds. Northcote and Skene (1972) defined sodic soils as those with ESP values between 6% and 14%, and strongly sodic soils as those with ESP >15%. Under the Australian Soil Classification (Isbell 1998), 'Sodosols' are identified at ESP >6%, although soils with ESP >25–30% are sometimes excluded due to severe structural degradation and land use constraints.

These discrepancies reflect a broader issue: the reliance on arbitrary chemical cut-offs rather than soil functional behaviour. Sumner *et al.* (1998) proposed shifting from purely compositional criteria (e.g. ESP or EC) to behaviour-based classification systems that consider properties such as dispersion, permeability, and structural resilience.

The impact of these inconsistencies is evident in global mapping efforts. Bui *et al.* (1998), using national and regional data, reported large variability in the

Table 2.1. World distribution of sodic soils (Bui *et al.* 1998)

Continent	Country	Area of sodic soils (×10^3 ha)
North America	Canada	6,974
	USA	2,590
South America	Argentina	53,139
	Bolivia	716
	Chile	3,642
Africa	Algeria	129
	Angola	86
	Botswana	670
	Cameroon	671
	Chad	5,950
	Ethiopia	425
	Ghana	118
	Kenya	448
	Liberia	44
	Madagascar	1,287
	Namibia	1,751
	Niger	1,389
	Nigeria	5,837
	Somalia	4,033
	Sudan	2,736
	Tanzania	583
	Zambia	863
	Zimbabwe	26
South Asia	Bangladesh	538
	India	574
	Iran	686
North and Central Asia	China	437
	USSR	119,628
Australasia	Australia	339,971

extent and severity of sodic soils across continents. Their estimates (summarised in Table 2.1) identified Australia as having the world's largest extent of sodic soils, followed by the former USSR. Bot *et al.* (2000) estimated that more than 400 million hectares globally were affected, a figure later revised to ~600 million hectares by the FAO and ITPS (2015). These numbers likely underestimate the true extent of sodic soils with dispersive behaviour, due to methodological limitations, inconsistent sampling depths, and the exclusion of structurally degraded soils falling below national ESP thresholds.

Collectively, these challenges highlight the need for globally consistent classification frameworks that integrate chemical indicators with functional soil behaviour, to better inform management and land use decisions.

2.5.2 Sodic dispersive soils in Australia

Australia, often described as the driest inhabited continent, has the world's most extensive area of sodic soils, estimated at ~340 million hectares (Table 2.2). These soils are largely products of long-term landscape evolution under late Cenozoic climatic regimes that favoured salt accumulation and the weathering of primary minerals. The most comprehensive national classification and mapping of salt-affected soils remains that of Northcote and Skene (1972), which, despite advances in technology, has not been comprehensively updated.

Sodic soils in Australia occur across diverse climatic zones but exhibit variable properties depending on rainfall, parent material, and landscape position. Approximately 86% are alkaline (pH >8.2), whereas 7% are acidic (pH <6). Classification is generally based on ESP values measured within the top 1 m of the soil profile as follows:

- ESP <6%: non-sodic
- ESP 6–15%: sodic
- ESP >15%: strongly sodic.

However, ESP alone does not reliably predict dispersive behaviour. Soil pH, mineralogy, and the presence of iron or aluminium oxides also influence structural stability. For example, acidic or near-neutral soils may not disperse at ESP >20% if aggregation is maintained by oxide coatings (Isbell 2002). A common feature of Australian sodic soils is their duplex character, comprising an acidic surface horizon overlying an alkaline, sodic subsoil. These sodic subsoils often act as hydraulic barriers, restricting water movement and causing seasonal waterlogging. In such systems, leaching from the surface horizon is

Table 2.2. Classification and area of salt-affected soils in Australia (Northcote and Skene 1972)

Note: 1,000,000 km^2 is equivalent to 100 Mha

	Salt-affected soil category	**Area (km^2)**	**% National land area**
Map unit			
SS	Saline soils	386,300	5.3
AS1	Alkaline strongly sodic to sodic clay soil with uniform structure	666,400	9.2
AS2	Alkaline strongly sodic to sodic coarse- and medium-textured soils	600,700	8.3
AS3	Alkaline strongly sodic to sodic duplex soils	454,400	6.3
NS1	Sodic and strongly sodic neutral duplex soils	134,700	1.9
NS2	Sodic acid duplex soils	140,700	1.9
Total		2,383,200	32.9

minimal, and salts can accumulate over time even in non-saline topsoils due to evapoconcentration.

Alkaline dispersive soils are especially prevalent in semi-arid and arid zones, where calcium and magnesium are preferentially leached, and sodium is retained. The high solubility of sodium carbonate (Na_2CO_3) and bicarbonate ($NaHCO_3$) relative to their calcium and magnesium analogues raises pH (often >8.5), promoting dispersion. In contrast, acid to neutral dispersive soils are more common in higher rainfall zones, where carbonate leaching reduces pH and base saturation, and dispersion is driven by different cation exchange dynamics (Rengasamy and Olsson 1991).

As shown in Table 2.2, sodic and potentially dispersive soils account for ~27.6% of Australia's land area, whereas all salt-affected categories combined cover 32.9%. These soils pose serious challenges for agriculture and land management due to their low permeability, structural instability, and limited plant-available water capacity. These constraints are particularly acute in dryland systems, where rainfall variability and soil–water interactions drive crop performance and land degradation.

Given the extent and diversity of sodic dispersive soils across the continent, region-specific diagnostics and management strategies are essential. Subsequent chapters explore targeted approaches to amelioration, focusing on physical, chemical, and biological interventions suited to different soil types and climatic conditions.

Salinity and sodicity remain among the most significant global threats to agricultural productivity and environmental stability. Despite their widespread impact, current global datasets on salt-affected and dispersive soils are often outdated, inconsistent, and heavily reliant on indirect indicators such as ESP and EC. There is a pressing need for more direct, behaviour-based diagnostics, particularly clay dispersion tests, to more accurately identify and map dispersive soils.

Climate change is expected to intensify these challenges. Higher temperatures will increase evapotranspiration and crop water demand, accelerating salt accumulation in surface soils. Coastal regions are particularly at risk, with rising sea levels and more frequent extreme weather events increasing seawater intrusion and the salinisation of soils and groundwater (Hopmans *et al.* 2022). In irrigated areas, the continued use of saline or marginal-quality water without adequate drainage infrastructure further compounds the risks of salinity and structural degradation.

To improve land management and rehabilitation outcomes, enhanced monitoring and classification systems are essential. Incorporating clay dispersion tests, together with concepts such as net dispersive charge and flocculating capacity, would greatly improve our ability to assess structural vulnerability and guide intervention strategies. More reliable and harmonised data are also needed to monitor trends, support regional reclamation efforts, and inform climate-resilient land use planning (Rengasamy *et al.* 2023).

The sustainable management of dispersive soils depends on accurate diagnosis, consistent classification, and responsiveness to climatic and hydrological dynamics. Without such an integrated approach, the global extent and severity of dispersive soils will likely continue to expand, posing serious consequences for food security, water quality, and long-term landscape resilience.

2.6 REFERENCES

Barrett-Lennard EG, Anderson CG, Holmes KW, Sinnott A (2016) High soil sodicity and alkalinity cause transient salinity in south-western Australia. *Soil Research* **54**, 407–417. doi:10.1071/SR15052

Bot AJ, Nachtergaele FO, Young A (2000) 'World soil resources report: land resource potential and constraints at 16 regional and country levels.' (Food and Agriculture Organization, Rome)

Bui EN, Krogh L, Lavado RS, Nachtergaele FO, Toth T, Fitzpatrick RW (1998) Distribution of sodic soils: The world scene. In 'Sodic soils. Distribution, properties, management, and environmental consequences'. (Eds ME Sumner, R Naidu) pp. 19–33. (Oxford University Press: New York, NY)

Emerson WW (1967) A classification of soil aggregates based on their coherence in water. *Soil Research* **5**, 47–57. doi:10.1071/SR9670047

FAO (1991) 'World soil resources report 66. An explanatory note on the FAO world soil resources map at 1:25 000 000 scale.' (Food and Agriculture Organization, Rome)

FAO and ITPS (2015) 'Status of the world's soil resources (SWSR) – main report.' (Food and Agriculture Organization of the United Nations and Intergovernmental Technical Panel on Soils, Rome)

Hillel D (2005) Soil salinity: historical and contemporary perspectives. In 'Proceedings of the International Salinity Forum – managing saline soils and water: science, technology and social issues – oral presentation abstracts'. pp. 231–234. (USDA-ARS Salinity Laboratory: Riverside, CA)

Holmes JW (1960) Water balance and the water table in deep sandy soils of the upper south-east Australia. *Australian Journal of Agricultural Research* **11**, 970–988. doi:10.1071/AR9600970

Hopmans JW, Qureshi AS, Kisekka I, Munns R, Grattan SR, Rengasamy P, *et al.* (2022) Critical knowledge gaps and research priorities in global soil salinity. *Advances in Agronomy* **169**, 1–191.

Isbell RF (1998) 'The Australian soil classification.' (CSIRO Publishing: Melbourne, Vic.)

Isbell RF (2002) 'The Australian soil classification.' (CSIRO Publishing: Melbourne, Vic.)

Jacobsen T, Adams RM (1958) Salt and silt in ancient Mesopotamian agriculture. *Science* **128**, 1251–1258. doi:10.1126/science.128.3334.1251

Kelley WP (1951) 'Alkali soils: Their formation, properties and reclamation.' (Reinhold Publishing, New York, NY)

Lawrie KC (2005) Salinity hazard and risk mapping: a multi-disciplinary approach for complex regolith landscapes in Australia. In 'Proceedings of the International Salinity Forum – managing saline soils and water: science, technology and social issues – oral presentation abstracts'. pp. 281–284. (USDA-ARS Salinity Laboratory: Riverside, CA)

Marchuk A, Rengasamy P (2012) Threshold electrolyte concentration and dispersive potential in relation to CROSS in dispersive soils. *Soil Research* **50**, 473–481. doi:10.1071/SR12135

Mohanavelu A, Naganna SR, Al-Ansari N (2021) Irrigation induced salinity and sodicity hazards on soil and groundwater: an overview of its causes, impacts and mitigation strategies. *Agriculture* **11**, 983. doi:10.3390/agriculture11100983

North PF (1976) Towards an absolute measurement of soil structural stability using ultrasound. *Journal of Soil Science* **27**, 451–459. doi:10.1111/j.1365-2389.1976.tb02014.x

Northcote KH, Skene JKM (1972) Australian soils with saline and sodic properties. CSIRO Soil Publication No. 27, CSIRO Divisions of Soils, Adelaide, SA.

Puri A, Keen B (1925) The dispersion of soil in water under various conditions. *Journal of Agricultural Science* **15**, 147–161. doi:10.1017/S0021859600005645

Rengasamy P (2002) Clay dispersion. In 'Soil physical measurement and interpretation for land evaluation'. (Eds N McKenzie, K Coughlan, H Cresswell) pp. 200–210. (CSIRO Publishing: Melbourne, Vic.)

Rengasamy P (2006) World salinization with emphasis on Australia. *Journal of Experimental Botany* **57**, 1017–1023. doi:10.1093/jxb/erj108

Rengasamy P (2010) Soil processes affecting crop production in salt-affected soils. *Functional Plant Biology* **37**, 613–620. doi:10.1071/FP09249

Rengasamy P, Olsson KA (1991) Sodicity and soil structure. *Soil Research* **29**, 935–952. doi:10.1071/SR9910935

Rengasamy P, Olsson KA (1993) Irrigation and sodicity. *Soil Research* **31**, 821–837. doi:10.1071/SR9930821

Rengasamy P, Tavakkoli E, McDonald GK (2016) Exchangeable cations and clay dispersion: net dispersive charge, a new concept for dispersive soils. *European Journal of Soil Science* **67**, 659–665. doi:10.1111/ejss.12369

Rengasamy P, McDonald GK, Tavakkoli E (2023) Managing multiple constraints to crop production in dispersive soils. In 'Soil constraints and productivity'. (Eds NS Bolan, MB Kirkham) pp. 319–341. (CRC Press, Boca Raton, FL)

Shainberg I, Shalhevet J (1984) 'Soil salinity under irrigation: processes and management.' (Springer-Verlag: Berlin)

Soil Survey Staff (1992) Keys to soil taxonomy. 5th edn. SMSS Technical Monograph No. 19. (Pocahontas Press: Blacksburg, VA)

Sumner ME, Rengasamy P, Naidu R (1998) Sodic soils: a reappraisal. In 'Sodic soils. Distribution, properties, management, and environmental consequences.' (Eds ME Sumner, R Naidu) pp. 3–17. (Oxford University Press: New York, NY)

Szabolcs I (1989) 'Salt-affected soils.' (CRC Press: Boca Raton, FL)

Wiebe BH, Eilers RG, Eilers WG, Brierley T (2005) Development of a risk indicator for dryland salinization on the Canadian Prairies. 'Proceedings of the International Salinity Forum – managing saline soils and water: science, technology and social issues – oral presentation abstracts'. pp. 473–476. (USDA-ARS Salinity Laboratory: Riverside, CA)

3

Chemical physics of dispersive soils

3.1 SOIL CHEMISTRY PRINCIPLES RELEVANT TO CLAY DISPERSION FROM SOIL AGGREGATES

Clay swelling and dispersion are fundamental processes underlying the structural instability of dispersive soils. Triggered by wetting, they lead to the breakdown of soil aggregates and result in poor soil physical performance. Although both processes arise from electrostatic and hydration forces between soil particles and the surrounding solution, they differ in outcome. For example, in saline soils, swelling may occur without dispersion due to the stabilising effect of high electrolyte concentrations.

Mechanistic understanding of clay behaviour in dispersive soils has traditionally been framed through models based on forces such as Lifshitz–van der Waals attraction, ion correlation, hydration energy, and diffuse double layer repulsion (e.g. Sumner 1993; Quirk 1994). However, these models were largely derived from studies of pure clay minerals suspended in aqueous media. Natural soils are far more complex: clay minerals are embedded in aggregates alongside sand, silt, organic matter, and oxides. Consequently, the behaviour of soil clays, confined within structured domains, cannot be directly equated to isolated clay particles in suspension.

Soil classification systems often label soils by dominant clay minerals, such as 'kaolinitic' or 'smectitic'. Yet, soil clays differ substantially from their mineralogical counterparts in terms of charge characteristics, bonding environments, and dispersion behaviour (Rengasamy and Sumner 1998). Therefore, understanding dispersion in natural soils requires moving beyond simplified colloid models to incorporate interactions among mineral and non-mineral phases within aggregate structures.

3.1.1 Soil structural stability of dispersive soils in water

Soil structure refers to the spatial arrangement of mineral and organic components into aggregates that define pore architecture and determine water and air flow. Structural stability is the ability of these aggregates to resist disintegration under mechanical or environmental stress. Aggregates form through a combination of physical, chemical, and biological processes that involve van der Waals forces, ionic and covalent bonds, hydrogen bonding, coordination complexes, and hydrophobic interactions (Rengasamy and Olsson 1991).

These aggregates exist in a hierarchical continuum: from floccules and domains to micro- and macroaggregates, and finally clods (Dexter 1988). Tillage and other mechanical disturbances can break aggregates apart if the applied stress exceeds the internal cohesive strength. However, in dispersive soils, even water alone can destabilise structure. Upon wetting, a process known as slaking may occur, followed by clay dispersion, especially when sodium dominates the exchange complex.

Although slaking alone may not always impair functionality, in dispersive soils it initiates a cascade where clay particles detach and remain in suspension. This collapse of structural hierarchy reduces porosity and impairs key functions such as infiltration and aeration. The resulting degradation manifests as crusting, compaction, and erosion, hallmarks of dispersive and sodic soils (Dexter 1988).

The initiation of swelling and dispersion is governed by interactions between water molecules and the electrostatic fields around charged particles. These interactions are influenced by the solvent's dielectric constant and the soil's exchangeable sodium percentage (ESP). In low-polarity solvents with small dielectric constants, slaking and dispersion are suppressed, even in soils with high ESP. Table 3.1 illustrates how solvents with varying dielectric properties influence slaking and clay dispersion in an Alfisol with contrasting sodicity.

Dispersion does not occur when no net surface charge is available for hydration. For example, heating Na-smectite above 300°C removes its swelling and dispersion

Table 3.1. Slaking and spontaneous clay dispersion of Alfisol aggregates (2–4 mm) at two exchangeable sodium percentage levels in various solvents (Rengasamy and Sumner 1998)

ESP, exchangeable sodium percentage

Solvent	Dielectric constant at 25°C	Dispersed clay (% of total clay)		Slaking (% of <2 mm)	
		ESP 1%	ESP 20%	ESP 1%	ESP 20%
Water	78.5	0	26	67	80
Ethanol	24.3	0	6	24	12
Benzene	2.3	0	0	0	0
n-Hexane	1.9	0	0	0	0

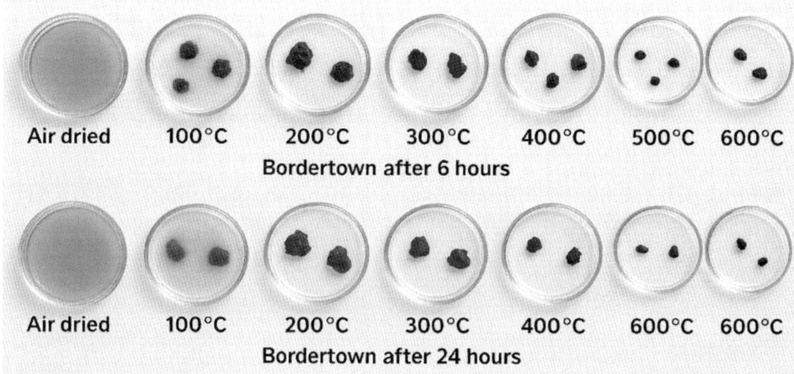

Fig. 3.1. Heating soil aggregates from Bordertown, South Australia (ESP ~17%), progressively reduces clay dispersion. Elevated temperatures induce thermal transformations that promote covalent bonding between sodium ions and clay minerals, thereby decreasing aggregate slaking and dispersion upon wetting.

capacity by altering surface bonding (Chorom and Rengasamy 1996). Similar results were observed in sodic Australian soils (ESP ≈17%), where aggregates became stable after thermal treatment (Fig. 3.1). These findings underscore that both the polarity of the wetting medium and the electrochemical properties of clay surfaces are critical in determining dispersion.

3.1.2 Processes leading to structural changes on wetting dry aggregates

Early explanations for swelling and dispersion relied on DLVO (Derjaguin–Landau–Verwey–Overbeek) theory (van Olphen 1977), which describes the interplay between attractive van der Waals forces and repulsive electrostatic interactions within the electrical double layer. In aqueous media, negatively charged clays are surrounded by hydrated counterions, and the resulting double layer governs particle stability (Rengasamy and Sumner 1998). Refinements to DLVO theory, such as the Gouy–Chapman and Stern models, aimed to better describe ion distributions and potential gradients (Gregory 1989; McBride 1994; Iwata *et al.* 1995).

However, DLVO theory has limitations when applied to natural soils. It does not account for ion-specific effects, the crystalline swelling of divalent-saturated clays under saline conditions, or the restricted development of diffuse layers within soil aggregates (Kjellander *et al.* 1988; Parsons *et al.* 2011). In soil systems, clay particles are not free colloids but are embedded in complex matrices of organic and inorganic constituents.

Understanding slaking and dispersion in these systems requires considering the sequence of hydration, swelling, and disintegration that occurs during wetting. The energy driving hydration depends on surface charge magnitude and accessibility, which are influenced by clay type, organic coatings, and inorganic associations. Not all surface charge is available for hydration; some is neutralised or shielded by organo–mineral interactions (Rengasamy and Olsson 1991).

For instance, in illite, potassium (K^+) forms strong inner-sphere complexes, limiting hydration, whereas smectite with sodium (Na^+) can form weaker outer-sphere complexes, promoting hydration and swelling (Slade *et al.*1991; Sposito *et al.* 1999). Upon initial wetting (Stage 1, Fig. 3.2), hydration forces reduce interparticle attraction, generating internal pressures in the megapascal (MPa) range. This leads to an interparticle spacing of 2–3 nm and a drop in aggregate strength from megapascal to kilopascal (kPa) levels (Stage 2).

In divalent cation-dominated systems, further swelling is constrained by stronger clay–cation bonding. For example, Mg-saturated clays swell more than Ca-saturated clays due to greater ionic character (Norrish 1954; Laird 2006), although both may undergo crystalline swelling, particularly under low ionic strength (Slade *et al.* 1991).

Slaking can occur when hydrostatic or pneumatic forces disintegrate weak aggregates. However, without prior electrostatic weakening, mechanical disturbance alone is often insufficient to cause dispersion. This is evident in soils wetted with non-polar solvents, where dispersion fails to occur despite high ESP (Rengasamy and Sumner 1998; Hu *et al.* 2018).

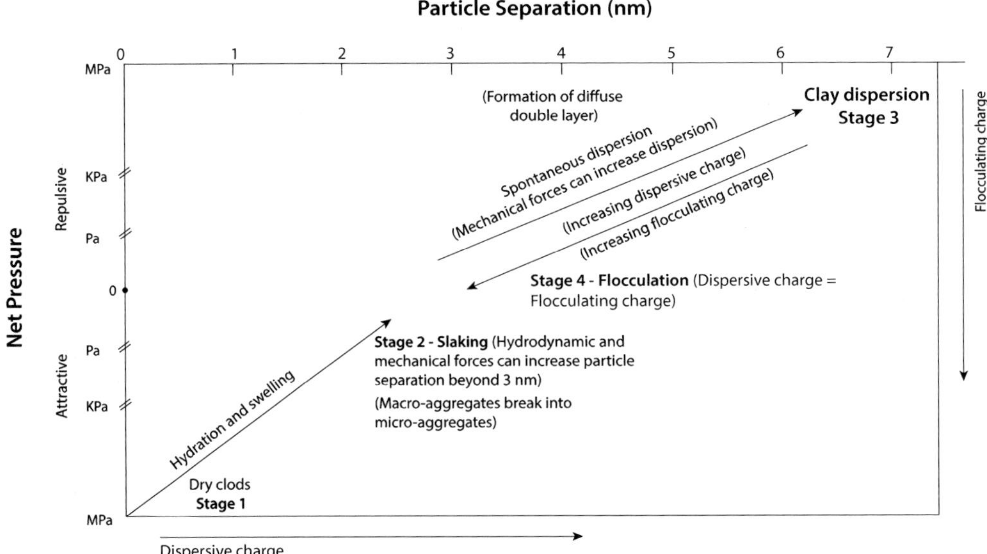

Fig. 3.2. Particle separation at different stages of wetting soil aggregates and net pressure generated by dispersive and flocculating charges (Rengasamy 2018).

As wetting progresses and monovalent cations (Na^+, K^+) dominate the exchange complex, swelling may exceed 7 nm, resulting in spontaneous dispersion (Stage 3). Clay particles become fully separated and remain suspended due to persistent repulsive hydration forces. In contrast, divalent-dominated systems resist spontaneous dispersion but may undergo mechanical dispersion if external forces, such as tillage or hydrological energy, are sufficient to destabilise aggregates (Emerson 1983; Emerson 2002).

Dispersion can also be enhanced by disruption of cementing agents (e.g. $CaCO_3$) or organo–mineral bonds, which may release fine particles into suspension. In the field, cultivation and rainfall can trigger this process in vulnerable soils (Rengasamy *et al.* 1984).

Flocculation occurs when repulsive forces are neutralised, either by increasing electrolyte concentration or adjusting pH to the point of zero charge (PZC), allowing van der Waals attraction to dominate (Stage 4). This transition is defined by the threshold electrolyte concentration, the minimum ionic strength needed to suppress dispersion (Chorom *et al.* 1994). Upon drying, attractive forces bring particles together, forming more stable aggregates.

This 'flocculation plus' process is strengthened by cation type and bonding configuration. If the water content remains below saturation and electrolyte levels exceed the threshold electrolyte concentration, swelling may occur without dispersion. However, leaching or dilution can reduce ionic strength below the flocculation threshold, reactivating dispersion, even in previously stable soils. These dynamic transitions highlight the complex interplay between water content, ionic strength, and cation composition in determining the stability or collapse of soil structure.

3.2 MECHANISMS OF CLAY DISPERSION FROM SOIL AGGREGATES

3.2.1 Repulsive forces in relation to cations and anions

The net surface charge on soil particles is governed by the mineralogy of clay minerals, associated inorganic constituents, and organic matter (Table 3.2). Although this charge is typically negative in most agricultural soils, it can be significantly reduced when components are linked by covalent bonding. According to thermodynamic principles, surface charges must be balanced by oppositely charged ions to maintain electrochemical stability.

In most soils, exchangeable cations, primarily Na^+, K^+, magnesium (Mg^{2+}), and calcium (Ca^{2+}), serve this balancing role. In acidic soils, additional contributions may come from hydrolysing cations such as aluminium (Al^{3+}), iron (Fe^{3+}), and manganese (Mn^{2+}), which can form covalent hydroxy complexes when bound to clay surfaces (Rengasamy and Oades 1979). In contrast, in highly weathered soils

Table 3.2. Soil factors controlling swelling, dispersion, and flocculation (after Rengasamy 2018)

Soil factors	**Mechanism**
Clay mineralogy and clay content	Charge originates in clay structures because of isomorphous substitution and broken bonds. The location of charge in tetrahedral structure is not available for hydration reactions. Thus, the total charge depends on the mineralogy and the amount of clay in soils
Soil pH	Adsorption of H^+ or OH^- ions alters the charge on broken bonds. With increasing concentration of carbonate anions, pH increases, and negative charge on soil particles increases. When pH decreases, as observed in acidic soils, negative charge decreases
Organic matter	Organic molecules bonded to clays by covalent bonding reduce the hydration charge of clay particles. Unbound, charged organic molecules can increase the hydration charge. Soil particles covered by hydrophobic organic matter are not affected by water interaction
Inner sphere complexes	Fixation of cations such as Fe^{2+}/Fe^{3+}, Al^{3+}, K^+ by clay minerals by inner sphere complexation (covalent bonding) reduces the hydration charge
Cementation	Cementation of soil particles by Fe and Al oxides or calcium carbonate can block the charge available for water interaction
Exchangeable cations	Exchangeable cations are attached to charged soil particles by a mixture of ionic and covalent bonding. The resultant ionicity of these bindings determines the net hydration charge. Dispersive charge depends on the dispersive power of the exchangeable cations
Electrolytes	Free (unbound) electrolytes in soil water contribute to the cationic flocculating charge, which is a function of the flocculating power and the concentration of individual cations

such as Oxisols (Ferrosols), where net positive charge may dominate, the charge is neutralised by exchangeable anions such as chloride, sulfate, and phosphate, highlighting the role of anion-controlled hydration and dispersion processes in these soils.

Traditionally, the interaction between exchangeable cations and clay particles was viewed as purely ionic. However, developments in inorganic chemistry show that most natural heteronuclear bonds exhibit both ionic and covalent character (Huheey *et al.* 1994). The degree of covalency is influenced by cation polarisability, determined by the extent to which a cation's electric field can distort an anion's electron cloud. This, in turn, is governed by the cation's charge-to-radius ratio (ionic potential) and the polarisability of the anion.

For instance, Ca^{2+} in $CaCl_2$ is more ionic and water soluble than Ca^{2+} in $CaCO_3$, reflecting greater covalent character and lower solubility in the latter. This relationship is significant for understanding how different bonding environments affect the hydration and dispersibility of clay–cation associations.

3.2.2 Development of ionicity and covalency indices for clay–cation bonds

The nature of clay–cation bonding is central to understanding dispersive behaviour. Polarisability, often approximated by the ionic potential ($IP = Z^+/r$, where Z^+ is the ionic charge and r is the ionic radius), reflects the ability of a cation to distort the electron cloud of the clay surface. Cations with higher polarisability tend to form more covalent (less hydrated) bonds.

Based on hard and soft Lewis acid theory, Misono *et al.* (1967) introduced a quantitative measure of cation softness, expressed as the Misono softness parameter (Y):

$$Y = 10(I_z/I_{z+1})(r_i/Z^{0.5})$$

where I_z and I_{z+1} are the ionisation potentials corresponding to successive oxidation states. Cations with low Y values are considered 'hard' (low polarisability), whereas those with high Y values are 'soft' (high polarisability; Sposito 2008). In the context of clay surfaces, which act as large, complex, high-charge anions, covalency increases when polarisable cations engage in inner sphere bonding.

To better quantify bonding behaviour, Marchuk and Rengasamy (2011) combined ionic potential and Misono's parameter to define a covalency index (CI):

$$CI = (I_z/I_{z+1}) \times Z^{0.5}$$

The corresponding ionicity index (II) is then calculated as:

$$II = 1 - CI$$

The CI reflects the covalent character of a cation's interaction with clays, whereas the II quantifies its susceptibility to hydration and dispersion. Typically, CI ≤ 1 and II values closer to 1 suggest greater ionic character and higher dispersive potential.

In geochemistry, a threshold IP of 30 nm^{-1} is often used to distinguish between non-hydrolysing (solvating) and hydrolysing cations (Sposito 2016). Cations such as Na^+, K^+, Mg^{2+}, and Ca^{2+}, with IP values below this threshold, do not hydrolyse extensively, whereas Al^{3+} and Fe^{3+}, with IPs of 76.9 and 46.2 nm^{-1} respectively, form hydroxo complexes with strong surface affinity.

The II is particularly valuable in predicting clay dispersion. Marchuk and Rengasamy (2011) demonstrated strong correlations between II and various measures of clay behaviour, including suspension turbidity, zeta potential, and mean particle size. As II increases, dispersion is more likely; conversely, higher CI values indicate stronger inner sphere bonding and greater structural stability. For example, the II ranks dispersive power as follows: $Li^+ > Na^+ > K^+ > Mg^{2+} > Ca^{2+} > Sr^{2+} > Ba^{2+}$.

Table 3.3. Ionisation potentials I_z and I_{z+1}, charge, and ionic radius (from the Handbook of Chemistry and Physics; Weast 1978) and the calculated covalent index and ionicity index of some common cations

I_z and I_{z+1} are ionisation potentials corresponding to successive oxidation states. CI, covalent index; II, iconicity index; r, ionic radius; Z, charge

Cation	Z	R (nm)	I_z (eV)	I_{z+1} (eV)	CI	II
Li	1	0.059	5.39	75.62	0.07	0.93
Na	1	0.102	5.14	47.29	0.11	0.89
K	1	0.138	4.34	31.81	0.14	0.86
Cs	1	0.167	3.89	25.10	0.15	0.85
Ag	1	0.115	7.57	21.48	0.35	0.65
Hg	1	0.119	10.43	18.75	0.56	0.44
Hg	2	0.102	18.75	34.2	0.78	0.22
Ca	2	0.100	11.87	51.21	0.33	0.67
Mg	2	0.072	15.03	80.14	0.27	0.73
Sr	2	0.118	11.03	43.60	0.36	0.64
Ba	2	0.135	10.00	35.50	0.40	0.60
Ni	2	0.069	18.15	35.16	0.73	0.27
Cu	2	0.057	20.29	36.83	0.78	0.22
Cd	2	0.095	16.91	37.48	0.64	0.36
Zn	2	0.074	17.96	39.70	0.64	0.36
Pb	2	0.119	15.03	31.93	0.67	0.33
Fe	2	0.078	16.18	30.64	0.75	0.25
Fe	3	0.065	30.64	56.80	1.00	0.00
Mn	2	0.083	15.63	33.69	0.66	0.34
Al	3	0.039	28.44	119.96	0.41	0.59
Cr	3	0.062	30.95	50.00	0.93	0.07
Si	4	0.026	45.13	166.73	0.54	0.46
Co	2	0.075	17.05	33.49	0.72	0.28
Co	3	0.061	33.49	83.10	0.70	0.30
As	3	0.058	28.35	50.13	0.98	0.02

These indices outperform both IP and Y values in predicting dispersive behaviour, offering a mechanistic basis for understanding how cation chemistry modulates soil structure.

The II also explains cation selectivity on negatively charged surfaces. Cations with high II (e.g. Cs^+, K^+) are held via electrostatic attraction and are easily exchangeable, whereas those with lower II (e.g. Al^{3+}, Fe^{3+}) tend to form surface complexes through covalent bonding (Sposito 2016). This pattern aligns with empirical selectivity sequences: $Cs^+ > K^+ > Na^+ > Li^+$ for monovalent cations and $Ba^{2+} > Sr^{2+} > Ca^{2+} > Mg^{2+}$ for divalent cations. These sequences are consistent with the II values listed in Table 3.3, supporting its use as a predictive tool in soil chemistry.

3.2.3 Exchangeable cations and clay dispersion: dispersive charge, flocculating charge and net dispersive charge

Research in Australia and elsewhere has demonstrated that clay dispersion is not driven by sodium alone; other exchangeable cations, particularly K^+ and Mg^{2+}, also influence soil structural stability. Emerson and Bakker (1973) showed that Na–Mg soils were more dispersive than equivalent Na–Ca soils. Curtin *et al.* (1994) quantified the dispersive effect of Mg as 4–5% of that caused by Na. Potassium similarly induces swelling and dispersion, with its effect reported to be similar to or slightly less than that of Na (Smiles and Smith 2004; Marchuk and Marchuk 2018).

Wilson *et al.* (2014) highlighted the role of clay surface charge and its pH sensitivity as key factors controlling dispersion. Rengasamy and Sumner (1998) concluded that the ionicity of clay–cation bonds determines their interaction with polar water molecules, leading to swelling and dispersion. This concept was further quantified by Marchuk and Rengasamy (2011), who developed IIs for common cations based on their physicochemical properties.

While ionicity reflects a cation's interaction with clay surfaces, the flocculation or dispersion of clays also depends on cation valence, as described by the Schulze–Hardy rule. Rengasamy and Sumner (1998) combined ionicity and valence effects to define the flocculating power of a cation as follows:

$$\text{Flocculating power} = 100 \times (I_z/I_{z+1})^2 \times Z^3$$

where Z is the valence, and I_z and I_{z+1} are the IPs for successive oxidation states (as introduced in Section 3.2.2). The calculated flocculating powers closely align with empirical data: relative to Na = 1, the values for K, Mg, and Ca are ~1.8, 27, and 45 respectively. Their dispersive powers, taken as the inverse of flocculating powers, are 45 for Na, 25 for K, and 1.88 for Mg (Rengasamy 2002*a*).

Building on these insights, Rengasamy *et al.* (2016) proposed a new concept that accounts for the dispersive and flocculating effects of all exchangeable and soluble cations influencing clay behaviour. This framework introduces the idea of dispersive charge, distinct from total cation exchange capacity (CEC) or layer charge.

It is critical to differentiate between the total layer charge of a clay mineral, resulting from isomorphous substitution, and the charge available for hydration. Illite, for example, has a higher layer charge than smectite but a lower CEC due to strong inner sphere complexation of K^+, which reduces hydration potential. Similarly, Na-saturated beidellite (with tetrahedral layer charge) tends to be more dispersive than montmorillonite (with octahedral charge) due to weaker cation–surface attraction (Slade *et al.* 1991).

The dispersive charge of a soil is also influenced by the nature of organic matter and inorganic cements. Covalently bound organic matter may reduce the hydration potential of clays, while cementing agents such as Fe and Al oxides or calcium carbonate can block reactive sites. Consequently, surface charge behaviour in natural soils cannot be predicted solely from bulk mineralogy or model clays (Bertsch and Seaman 1999). Soil pH further modulates net charge; as pH changes, the degree of dissociation or protonation of functional groups alters the charge available for interaction with water (Chorom and Rengasamy 1995).

In most agricultural soils with net negative charge, dispersion is primarily influenced by exchangeable cations. In contrast, in positively charged oxidic soils (e.g. tropical Oxisols), dispersion may be influenced by exchangeable anions. The key determinant of dispersive behaviour remains the net ionicity of the clay–cation bond: the higher the ionicity, the greater the water–clay interaction and potential for dispersion. The ionicity order for common cations is: Na > K > Mg > Ca (Rengasamy and Marchuk 2011). Their dispersive powers, inversely related to flocculating strength, follow the same ranking. Relative to Ca = 1, Mg, K, and Na have dispersive powers of 1.7, 25, and 45 respectively (Rengasamy 2002*a*). Importantly, both the concentration and the dispersive power of exchangeable cations influence the overall degree of clay dispersion.

To quantify these effects, Rengasamy *et al.* (2016) defined the dispersive charge of a soil as:

$$\text{Dispersive charge} = (\text{Ca}) + 1.7(\text{Mg}) + 25(\text{K}) + 45(\text{Na})$$

where each cation's concentration (in $\text{cmol}_c\ \text{kg}^{-1}$) is measured from the exchange phase at a given soil pH.

In contrast, the flocculating charge is derived from the concentration of soluble (unbound) cations in the soil solution and reflects their capacity to suppress dispersion:

$$\text{Flocculating charge} = 45(\text{Ca}) + 27(\text{Mg}) + 1.8(\text{K}) + (\text{Na})$$

Both dispersive and flocculating charges are calculated by applying empirically derived weighting factors (from Rengasamy and Sumner 1998) to cation concentrations. The balance between these opposing effects defines the net dispersive charge, which determines the extent of clay dispersion in a suspension, as follows:

$$\text{Net dispersive charge} = \text{Dispersive charge} - \text{Flocculating charge}$$

At the zero point of dispersion, dispersive and flocculating charges are equal, and complete flocculation occurs. As with surface charge, soil pH influences the net particle charge and the PZC, which is the pH at which net ionic adsorption is neutral. At the PZC, soils neither swell nor disperse, regardless of cation concentration or composition.

Together, these concepts offer a mechanistic framework for predicting clay behaviour based on the interplay of exchangeable cations, solution chemistry, ionicity, and soil pH, advancing beyond ESP alone to a more nuanced understanding of dispersive soils.

3.2.4 Concept of 'dispersive potential'

To better capture the combined effects of exchangeable sodium and electrolyte concentration on soil structural behaviour, Rengasamy *et al.* (1991) introduced the concept of dispersive potential (P_{DIS}). This parameter quantifies the energy imbalance that leads to clay dispersion and is defined as the difference between the threshold electrolyte concentration (the concentration required to prevent dispersion (P_{TEC})) and the ambient soil solution concentration (P_{SOL}), expressed in osmotic pressure units:

$$P_{DIS} = P_{TEC} - P_{SOL}, \text{ for } P_{SOL} < P_{TEC}$$

The methodology for estimating these parameters, including how to weight divalent cations using the Schulze–Hardy rule, was outlined in Rengasamy (2002*a*). Empirical studies showed that P_{DIS} was associated with key sodic soil properties such as aggregate stability, clay dispersion, and crusting strength. However, later research highlighted some limitations. Specifically, the effect of soil pH on net surface charge and the inapplicability of the Schulze–Hardy rule to natural clay systems complicated the interpretation of P_{DIS} (Rengasamy and Sumner 1998).

Subsequent advances in understanding clay–cation interactions, particularly through quantification of cation ionicity, dispersive charge, and flocculating charge, led to a refinement of the concept. Rengasamy *et al.* (2016) proposed the more robust and mechanistic concept of net dispersive charge, defined as:

$$\text{Net dispersive charge} = \text{Dispersive charge} - \text{Flocculating charge}$$

This model integrates both the dispersive influence of exchangeable cations and the flocculating effect of soluble cations in the soil solution. It accounts for differences in ionicity and valence, and offers a more realistic basis for predicting clay behaviour.

Using this framework, Rengasamy *et al.* (2016) analysed 96 soils from southern Australia and found that net dispersive charge was strongly correlated with the measured quantity of dispersed clay (Pearson's $r = 0.89$; Fig. 3.3). In that study, CEC was measured at pH 7.0; however, most soils were alkaline (pH 8.5–9.5). Because CEC and cation exchange values vary with pH, recalculating dispersive charge at the actual soil pH would likely strengthen this correlation. Supporting this model, Farahani *et al.* (2019) also reported strong correlations between net dispersive charge, dispersed clay, and zeta potential. They additionally observed

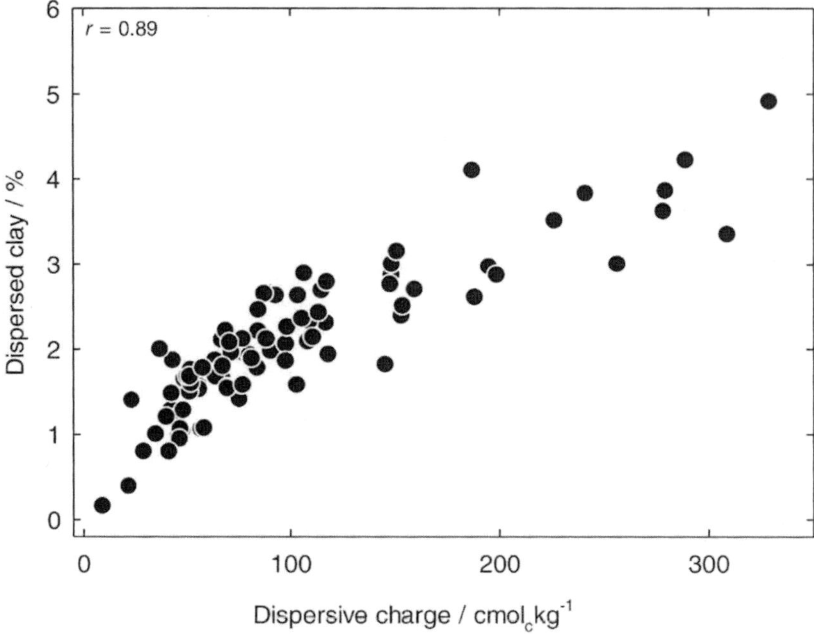

Fig. 3.3. Relationship between net dispersive charge ($cmol_c$ kg^{-1}) and dispersed clay (%) across 96 soil samples collected from southern Australia. Each point represents the mean of three replicates (n = 3). A strong positive correlation indicates that increasing net dispersive charge is associated with greater clay dispersion. Adapted from Rengasamy *et al.* (2016).

that the exchange coefficient for Na^+ and K^+ depended on the ionicity of the cation–clay bond, reinforcing the predictive value of ionicity indices in determining clay behaviour.

The concept of net dispersive charge also helps resolve long-standing controversies in the literature regarding the roles of K^+ and Mg^{2+} in dispersion and soil structural degradation. Because dispersive charge is calculated based on the charge available for hydration, it inherently incorporates the effects of clay mineralogy, organic matter, exchangeable cations, and pH. Simultaneously, the flocculating charge integrates the effects of electrolyte concentration and cation type, eliminating the need to rely solely on electrical conductivity (EC) as a proxy for stability. Indeed, EC provides a measure of total ionic concentration, but not flocculating capacity. For example, a soil solution with an EC of 1 dS m^{-1} reflects an electrolyte concentration of ~10 $mmol_c$ L^{-1}. If this concentration consists entirely of Ca^{2+}, its flocculating power is equivalent to 450 $mmol_c$ L^{-1} of Na^+, which would have an EC of ~45 dS m^{-1}. Thus, flocculating charge depends not just on total ionic strength but also on the specific composition and flocculating power of the cations present.

In summary, the net dispersive charge model integrates both the chemical and physical principles of dispersion, offering a predictive, mechanistic, and scalable tool for assessing soil structural vulnerability in sodic and dispersive landscapes.

3.2.5 Mechanism of flocculation of soil aggregates in dispersive soils

Just as the classical double layer (DLVO) theory fails to adequately explain the full complexity of clay dispersion in natural soils, it also falls short in describing flocculation behaviour, particularly in dispersive soils. DLVO theory assumes uniform electrostatic and van der Waals interactions in ideal colloidal suspensions, but it does not account for ion-specific effects, especially the differential flocculating powers of cations based on their ionicity and valence. In natural dispersive soils, flocculation (the stabilisation of soil aggregates) is better understood as the prevention of clay dispersion. It occurs when the flocculating charge, primarily contributed by electrolytes in the soil solution, exceeds the dispersive charge arising from the exchangeable cations on clay surfaces. When this balance favours flocculation, clay particles remain aggregated during wetting and dispersion is suppressed.

In saline soils with high concentrations of electrolytes, the ionic repulsion that drives dispersion is neutralised. High electrolyte concentrations reduce the thickness of the diffuse double layer and draw particles closer, increasing attractive forces. Moreover, cations with higher covalent bonding potential (i.e. a lower II) such as Ca^{2+} or Mg^{2+} are more effective at stabilising clay particles than monovalent cations like Na^{+} or K^{+}. This stabilisation effect is not simply due to total ionic strength, but to the specific flocculating power of each cation, governed by its valence, size, and ionicity (Fig. 3.2; Table 3.3).

For example, in saline systems dominated by low-flocculating-power monovalent cations, dispersion may still be prevented due to osmotic effects alone. High osmotic pressure compresses the diffuse double layer and reduces particle separation, increasing interparticle attraction. This effect, although not dependent on cation valency, limits dispersion under saline conditions. Conversely, in dispersive soils with low electrolyte concentrations, the addition of flocculating cations with high bonding strength, such as Ca^{2+} from gypsum, suppresses dispersion by neutralising the negative surface charge through inner sphere complexation and electrostatic attraction. These cations replace weaker, more dispersive cations like Na^{+}, stabilising the soil structure and reducing clay turbidity in suspension.

These data clearly demonstrate the relative efficiency of different cations in promoting clay flocculation. Divalent cations such as Ca^{2+} and Mg^{2+} induce flocculation at significantly lower concentrations than monovalent cations like Na^{+} and K^{+}. The relative flocculation concentration (RFC) is defined as the ratio of

Table 3.4. Critical flocculation concentrations (CFC) for homo-ionic clays from an Australian Alfisol in electrolytes with same cations and chloride anion (after Rengasamy and Sumner 1998)

The relative flocculation concentration (RFC) is calculated as CFC_{XCl}/CFC_{CaCl2}, where CFC is the critical flocculation concentration and X is Na, K or Mg. Mean RFC shows data averaged over four types of soil clays reported by Rengasamy and Sumner (1998)

Clay form	Electrolyte	CFC ($mmol_c$ L^{-1})	RFC	Mean RFC
Na-clay	NaCl	14.0	46.7	44.8
K-clay	KCl	8.5	28.3	27.3
Mg-clay	$MgCl_2$	0.6	2.0	1.75
Ca-Clay	$CaCl_2$	0.3	1.0	1.00

the critical flocculation concentration (CFC) of a given cation to that of Ca^{2+}. An RFC value of 1.0 denotes equivalent flocculating power to Ca^{2+}, whereas higher RFC values indicate reduced effectiveness (Table 3.4).

In acidic dispersive soils, additional multivalent cations, particularly Mn^{2+}, Al^{3+}, and Fe^{3+}, may also influence clay flocculation. However, their behaviour in soil solution is considerably more complex, as they readily hydrolyse to form a variety of hydroxy complexes whose net charge varies with pH. This complexity often leads to unpredictable interactions with soil colloids. For instance, Jenkins and Morand (2004), studying acidic (pH <5.5) sodic soils in New South Wales, reported highly variable relationships among clay dispersion, ESP, CEC, and aluminium concentrations, with no consistent or interpretable trends.

Further evidence of this complexity was observed in preliminary experiments by T. R. de Melo, I. Andelkovic, and P. Rengasamy (pers. comm.). In these trials, soil clay suspensions were treated with $FeCl_3$ and $AlCl_3$ solutions across a range of pH levels and compared to Ca^{2+}-induced flocculation. Although both Al^{3+} and Fe^{3+} were capable of inducing flocculation, their effective concentrations varied with pH and did not follow any consistent pattern relative to Ca^{2+} (Table 3.5). This irregular behaviour underscores the intricate speciation and hydrolysis dynamics of Al and Fe in soil systems, and highlights the need for detailed mechanistic studies on their surface reactivity and interaction with clay minerals. As shown in Table 3.5, the flocculating concentrations of Al^{3+} and Fe^{3+} varied with pH, but did not exhibit a consistent pattern relative to Ca^{2+}. This irregularity highlights the complex speciation of Al and Fe in soil systems, and the need for detailed studies on their hydrolysis products and surface reactivity.

At lower pH, Fe^{3+} and Al^{3+} can act as effective flocculants, possibly due to the formation of hydroxy cations with high positive charge density. However, as pH increases, their speciation becomes more complex, and their effectiveness less predictable. In contrast, Ca^{2+} maintains consistent flocculating behaviour across a wider pH range.

Table 3.5. Concentrations of Al^{3+} and Fe^{3+}, compared with Ca^{2+}, required to flocculate a soil clay

				Flocculating concentration relative to Ca	
pH	Al^{3+} ($mmol_c\ L^{-1}$)	Fe^{3+}($mmol_c\ L^{-1}$)	Ca^{2+}($mmol_c\ L^{-1}$)	Al/Ca	Fe/Ca
5.0	1.8	4.6	15.0	0.1	0.3
5.8	5.9	13.1	26.0	0.2	0.5
6.5	10.7	18.0	26.5	0.4	0.7
7.5	7.8	14.5	34.8	0.2	0.4
8.0	12.0	16.0	35.5	0.3	0.5

In conclusion, the mechanism of flocculation in dispersive soils involves more than electrolyte concentration alone. It is governed by the cationic composition, bonding strength (ionicity), valence, and pH-dependent speciation, especially for multivalent cations like Al^{3+} and Fe^{3+}. Although flocculation is traditionally defined as the aggregation of suspended particles, in soil systems it is best understood as the prevention of clay dispersion: a dynamic balance between repulsive and attractive forces mediated by the soil's chemical environment.

3.3 PHYSICAL BEHAVIOUR OF DISPERSIVE SOILS

The physical degradation of dispersive soils has been widely documented in numerous reviews (e.g. Shainberg and Letey 1984; Gupta and Abrol 1990; Rengasamy and Olsson 1991; So and Aylmore 1993; Qadir and Schubert 2002). Studies have consistently demonstrated that increasing exchangeable ESP results in clay dispersion, leading to soil crusting, hard-setting, increased soil strength, and erosion, while simultaneously reducing saturated and unsaturated hydraulic conductivity, infiltration, drainage, aeration, and plant-available water retention.

3.3.1 Structural degradation

Regardless of the dominant exchangeable cations, dispersive soils are characterised by degraded structure, which represents a major limitation to crop production. Although salinity is often low, meaning osmotic effects on plant growth are minimal, nutrient deficiencies may arise due to restricted ion mobility or reduced root access to water.

Structural degradation impairs water infiltration, oxygen diffusion, temperature regulation, and biological activity. These conditions collectively hinder seedling emergence, root growth, and the uptake of water and nutrients. So and Aylmore (1993) proposed a conceptual model linking dispersion to physical deterioration and yield loss. Dispersive topsoils are especially prone to waterlogging

Table 3.6. Flocculating power (relative to Na^+) and dispersive power (relative to Ca^{2+}) of cations commonly found in soils in relation to the ionicity index

Cation	Ionicity index	Flocculating power	Dispersive power
Na^+	0.89	1.0	45.0
K^+	0.86	1.8	25.0
Mg^{2+}	0.73	27.0	1.6
Ca^{2+}	0.67	45.0	1.0

Table 3.7. Active porosity, saturated hydraulic conductivity, and zeta potential of the dispersed clay in relation to the degree of clay dispersion of an Alfisol treated with different cations (after Marchuk *et al.* 2012)

NTU, nephelometric turbidity unit

Cationic form	Dispersive charge ($cmol_c$ kg^{-1})	Turbidity (NTU)	Dispersed clay (%)	Zeta potential (mV)	Active porosity (%)	Hydraulic conductivity (cm h^{-1})
Na^+	1,620	2,800	4.5	-50.1	2.6	0.08
K^+	900	1,540	2.8	-44.8	8.4	0.14
Mg^{2+}	58	970	0.3	-26.2	17.9	0.92
Ca^{2+}	36	500	0.1	-22.0	38.6	1.00

during moderate rainfall events and form hard crusts upon drying, reducing germination rates. Although subsoil layers may not appear dispersive under dry conditions, saturation can activate dispersive reactions. As clays detach and migrate into pore spaces, they clog channels, increase compaction, and further restrict hydraulic and gaseous conductivity. These processes limit soil water storage and root penetration, especially in dryland cropping systems.

3.3.2 Clay dispersivity and physical properties in relation to exchangeable cations

The physical behaviour of dispersive soils is directly linked to the ionicity of clay–cation bonds. As discussed in Section 3.2, cations with higher ionicity contribute to greater dispersive charge and structural breakdown. Table 3.6 summarises the flocculating and dispersive powers of key soil cations relative to Na^+ and Ca^{2+} respectively. Sodium exhibits the highest dispersive power, followed by K^+, Mg^{2+}, and Ca^{2+}.

The extent of clay dispersion governs key physical properties, such as active porosity and hydraulic conductivity. Increasing dispersion reduces porosity and water movement (So and Aylmore 1993; Marchuk *et al.* 2012). Transmission electron microscopy has revealed distinct microstructural changes in cation-treated clays, with increased aggregation under Ca^{2+} and Mg^{2+} and more dispersed structures under Na^+ and K^+ (Table 3.7; Fig. 3.4).

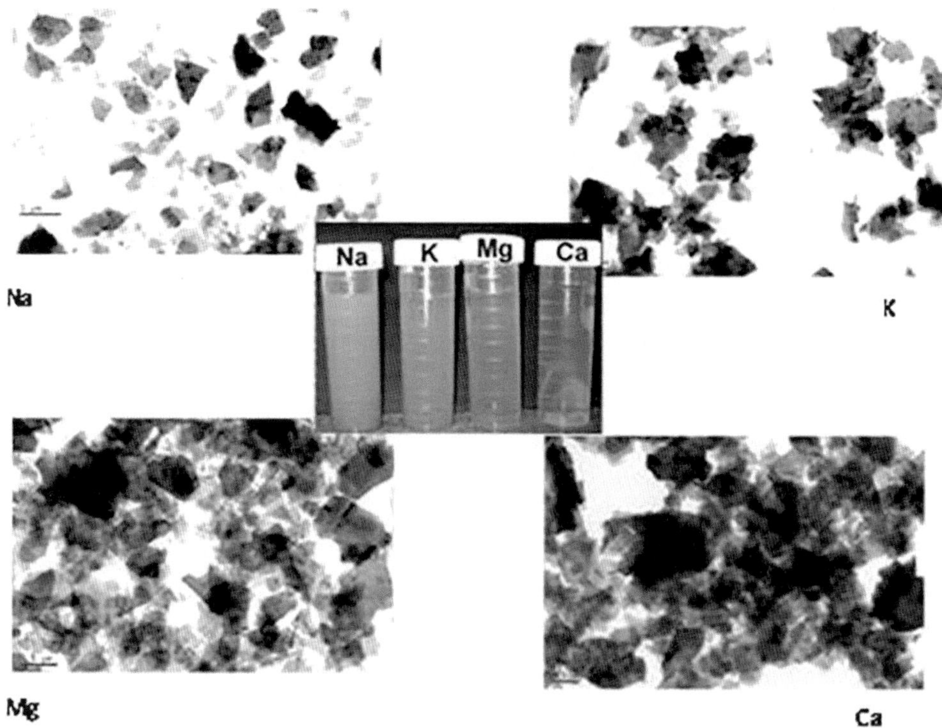

Fig. 3.4. Transmission electron micrographs of dispersed clay particles from homoionic Na^+, K^+, Mg^{2+}, and Ca^{2+} forms of an Alfisol soil. The images highlight the morphological differences in clay dispersion and particle association under the influence of different cations. Adapted from Marchuk and Rengasamy (2011).

Anion effects are also important: chloride and sulfate typically stabilise soil pH around neutral, whereas bicarbonate and carbonate increase pH above 8.5, thereby promoting clay dispersion (Tavakkoli *et al.* 2015). Importantly, only exchangeable K^+ contributes to dispersion; fixed K^+ in inner sphere complexes does not (Marchuk and Marchuk 2018).

Although Li^+ is not common in most soils, its high ionicity means it can contribute significantly to dispersion if introduced through mining activities (Marchuk and Rengasamy 2011). Among divalent cations, Mg^{2+} causes slightly more dispersion than Ca^{2+}, but both are far less dispersive than monovalent cations. Curtin *et al.* (1994) estimated Mg^{2+} has ~5% of the dispersive effect of Na^+ at comparable concentrations.

With increasing global use of recycled and saline waters, the impact of elevated Mg^{2+} concentrations on soil structural integrity has become a concern (Qadir *et al.* 2018). Therefore, understanding the combined effects of exchangeable cations, salinity, and pH is essential for evaluating the structural vulnerability of soils.

3.3.3 Water use efficiency in dispersive soils

In the context of Australian agriculture, efficient water use is critical due to limited availability of 'blue water' (from rainfall, rivers, and groundwater) and increasing reliance on 'grey water' (recycled or treated wastewater). However, 'green water' (soil-stored water available to crops) remains the primary water source for dryland farming systems.

Of the 120,000 GL of annual rainfall received across Australia, ~95,700 GL is temporarily stored in soils. Yet, an estimated 28,710 GL remains unused by crops, largely due to soil structural constraints and associated abiotic stress (P. Rengasamy, unpubl. obs.). Poor soil structure in dispersive soils prevents adequate root access to this stored water, diminishing water use efficiency.

So and Aylmore (1993) demonstrated that increasing clay dispersion correlates with higher soil strength and lower values of saturated hydraulic conductivity, water retention, and plant-available water. These relationships directly impact crop yields, particularly in dryland systems where irrigation is not feasible.

Figure 3.5 illustrates the concept of non-limiting water range, the range of soil water content within which plant growth is not constrained by aeration or mechanical resistance (Letey 1985). In dispersive soils, this range is significantly

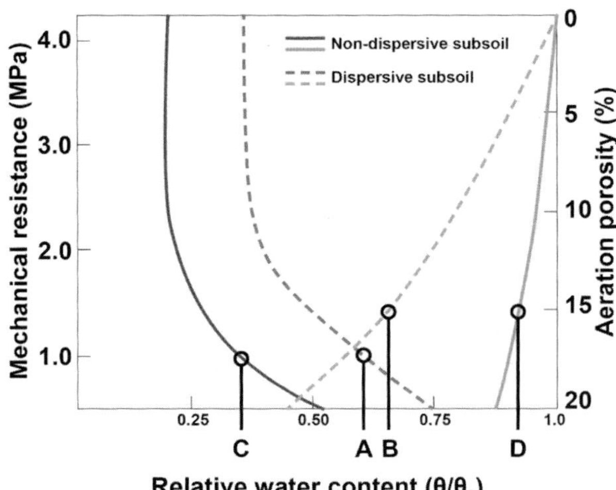

Fig. 3.5. Non-limiting water range in a dispersive soil (points A to B) compared with a non-dispersive soil (points C to D) in relation to mechanical strength and aeration porosity from the data of Rengasamy *et al.* (1992). The dark dashed and dark solid lines on the left show the increase in soil strength as drying occurs in the dispersive and non-dispersive soils respectively. The grey dashed and grey solid lines on the right show the decrease in air-filled porosity as wetting occurs in the dispersive and non-dispersive soils respectively.

narrower than in non-dispersive soils. In Fig. 3.5, point 'B' represents the water content above which aeration porosity exceeds 15%, whereas point 'A' represents the water content below which mechanical resistance drops below 1 MPa. The narrow interval between points B and A reflects the limited conditions under which plant roots can function optimally in dispersive soils.

3.3.4 Soil strength in relation to dispersivity

Dispersive soils become dense, cloddy, and structureless upon drying due to the destruction of natural aggregation. At the surface, dispersed clay particles form hard crusts, typically up to 10 mm thick, that act as cementing agents. These crusts hinder seedling emergence and may tear roots as they dry and shrink. The degree of crusting depends on texture, clay mineralogy, the extent of dispersion, and the rate of drying. Soils high in montmorillonite content often crack on drying due to their expansive properties.

Crusting affects only the upper few millimetres, whereas hard-setting soils undergo full aggregate breakdown throughout the horizon. In hard-setting soils, dispersed clays move vertically within the profile, sealing pores and reinforcing high soil strength. Modulus of rupture and penetrometer resistance are commonly used to assess strength, especially under field-relevant moisture conditions. Studies show that increasing dispersive potential and dispersed clay content correlates strongly with increased soil strength and resistance (Rengasamy *et al.* 1993; So and Aylmore 1993; Barzegar *et al.* 1994).

3.3.5 Transient salinity in dispersive soils

In regions with deep water tables and restricted leaching, salts accumulate within the soil profile from rainfall, weathering, and Aeolian inputs (see Fig. 2.2). This seasonal and spatially variable form of salinity, referred to as transient salinity (Rengasamy 2002*b*), is common in dispersive and sodic soils. Often mislabelled as 'dry saline land' or 'magnesia patches' in South Australia, the dominant salts are typically sodium-based.

Transient salinity affects ~33% of Australia's land area (Rengasamy 2002*b*; Barrett-Lennard *et al.* 2016). Salt accumulation may occur in surface or subsurface layers depending on hydraulic conditions and alters soil classification: a dispersive soil may become saline–dispersive or saline depending on the salt type and concentration. While dispersive physical constraints persist, the additional osmotic and ionic stresses vary with salt composition and plant species sensitivity.

If dominated by monovalent salts, saline–dispersive soils can revert to dispersive status after amendment and leaching. For example, gypsum application enhances infiltration and leaches sodium salts, thereby reducing both osmotic stress and dispersion potential.

3.4 HYDROLOGY AND EROSION

Increased clay dispersion compromises hydraulic conductivity by blocking soil pores in both surface and subsurface layers. This leads to surface sealing, waterlogging, and the development of impermeable subsoil horizons. Water stagnation promotes surface run-off, soil erosion, and subsoil instability, including tunnelling and piping (Stumpf 2013). During rainfall or irrigation, dispersed colloids are easily transported off-site, reducing topsoil fertility and contaminating waterways.

3.4.1 Waterlogging in dispersive soils

Waterlogging is common in dispersive soils due to low porosity, poor structure, and restricted infiltration. Dispersive soils, particularly those with waterlogging issues, often exhibit strong texture contrasts between sandy topsoils and dispersive clay subsoils. Infiltration is initially rapid but slows dramatically upon reaching the compacted subsoil, causing perching of water and saturating the root zone. This creates anaerobic conditions that reduce oxygen diffusion, suppress microbial activity, and impair plant root function. Excess soil water limits gaseous exchange, resulting in oxygen deficiency (anoxia), which can cause root death, nutrient imbalances, and reduced growth. These conditions are especially problematic in alkaline dispersive soils (Rengasamy *et al.* 2022).

Waterlogging is typically assessed using shallow piezometers or through the sum of excess water at 30 cm soil depth (SEW30), which integrates the number of days and the depth by which water tables rise above 30 cm (Setter and Waters 2003). SEW30 values below 30 cm·days indicate well-drained soils, while values above 1,200 or 2,500 cm·days reflect poor or very poor drainage respectively (Setter and Waters 2003).

3.4.2 Erosion

Clay dispersion is a major driver of erosion in dispersive soils, increasing vulnerability to rill, interrill, and gully erosion (Sumner *et al.* 1998). Interrill erosion occurs when dispersed particles are detached and transported across the soil surface by thin sheets of water. Rill erosion involves concentrated flow paths, which can rapidly deepen and widen, evolving into gullies. Gullies often initiate piping or tunnelling, particularly in dispersive subsoils (Boucher 1990; Hardie *et al.* 2009).

Tunnel erosion is a hidden yet destructive process where the subsurface flow of water mobilises dispersed clays into a slurry. This slurry follows natural conduits, such as surface cracks, root channels, or animal burrows, and, if a sufficient gradient exists, flows downslope beneath the surface. Over time, tunnels expand, eventually leading to roof collapse, potholes, and open gullies (Hardie *et al.* 2009). Tunnel erosion frequently goes undetected until major surface damage occurs.

Severe sheet and gully erosion associated with dispersive sodic soils has been reported across Australia, Canada, New Zealand, Papua New Guinea, the US, and Zimbabwe (Boucher 1990). According to Goldsmith and Smith (1985), five conditions are necessary for tunnel formation:

- a source of water (e.g. rainfall or irrigation)
- an infiltration rate that exceeds subsoil permeability
- a dispersive layer overlying a lower permeability layer
- a hydraulic gradient to induce lateral water movement
- an outlet to allow flow from the subsurface.

Tunnel erosion is often triggered by surface run-off entering the soil via desiccation cracks or root channels and encountering a dispersive, poorly permeable layer. Once initiated, these tunnels propagate rapidly under wet conditions, forming highly unstable networks and eventually contributing to landscape degradation.

3.5 SOIL MECHANICS AND ENGINEERING PROPERTIES

Dispersive soils present significant challenges in geotechnical engineering and infrastructure development. Their tendency to undergo clay dispersion and associated structural collapse can adversely affect the performance and longevity of civil engineering works, including buildings, roads, earth-filled dams, irrigation canals, and embankments (Sherard *et al.* 1977; Faulkner 2006).

One of the most critical geotechnical concerns in dispersive soils is piping failure in earth dams and embankments. Piping occurs when water flows through a concentrated leakage path, such as a shrinkage crack or poorly compacted zone. In dispersive soils, this water movement leads to rapid deflocculation and erosion of clay particles along the channel walls. Once initiated, erosion progresses along the entire length of the flow path simultaneously, rather than propagating from a single point. Such failures are especially common in areas with high cracking potential, including zones near conduits, regions of compressibility contrast, and locations affected by desiccation (Sherard *et al.* 1977).

The structural instability of dispersive soils stems from two main physical processes: (1) the breakdown of particle aggregates during wetting; and (2) the swelling of clays in response to moisture changes. Both mechanisms reduce the shear strength of the soil, compromising its ability to support load and resist deformation. As a result, infrastructure built on or within dispersive soils is at elevated risk of deformation, settlement, or catastrophic failure (Baldermann *et al.* 2021).

In cold or seasonally frozen regions, landslide hazards are compounded by freeze–thaw cycles and freeze-induced soil disturbance. Dispersive soils in these environments are particularly susceptible to both shallow-seated and deep-seated landslides. Shallow-seated failures often initiate when dispersion and structural collapse reduce the strength of near-surface soils, whereas deep-seated landslides may develop following prolonged rainfall infiltration. This infiltration leads to further erosion, the development of tensile cracks near the crest of slopes, and ultimately slope failure (Wang *et al.* 2020).

The triggering of landslides or embankment failure is consistently linked to the presence of water and pre-existing structural weaknesses, such as shrinkage cracks, construction deficiencies, or heterogeneities in soil properties (Parameswaran 2016). In addition to structural erosion, dispersive soils may exhibit differential settlement and poor compaction characteristics, further compromising engineering performance.

Given these risks, it is essential that soil dispersion potential is properly diagnosed during the design and construction phases of infrastructure projects, particularly those involving earthworks or hydraulic engineering. Where dispersive soils are present, mitigation measures such as chemical stabilisation (e.g. gypsum or lime), improved compaction, and erosion control structures must be carefully considered.

3.5.1 Soil strength and bearing capacity

The bearing capacity of a soil refers to its ability to support structural loads without undergoing shear failure. Bearing capacity depends on soil type, shear strength, and density. Shear strength determines the ultimate load a soil can carry and dictates slope stability in road cuts and embankments (Brink *et al.* 1984). When dispersive clays swell and shrink with changing moisture conditions, the resulting soil movement can damage road pavements, foundations, and underground utilities such as water pipes and optic cables (Palmer and Hazelton 1994).

Road infrastructure requires a stable subgrade with low dispersivity and erosion risk. Structural elements such as culverts and bridge footings are particularly sensitive to the erosion of dispersive materials around them. Where stable soils (e.g. compacted gravel) offer bearing capacities of ~440 kPa, soft dispersive clays may exhibit values below 75 kPa, making them unsuitable for supporting heavy infrastructure (Biggs and Mahony 2004). The required mechanical performance of road pavements (i.e. strength, durability, impermeability, wear resistance, and workability) is compromised in dispersive soils due to aggregate breakdown and low cohesion. Thus, geotechnical site investigations must consider dispersive behaviour when evaluating foundation or pavement design in affected regions.

3.5.2 Plasticity and Atterberg limits

The Atterberg limits, specifically the plastic limit and liquid limit, define the moisture content at which a soil transitions between solid, plastic, and liquid states. These limits provide insight into soil consistency and are routinely used to infer properties such as compressibility, shrink–swell potential, hydraulic conductivity, and shear strength.

The plasticity index, calculated as the difference between the liquid limit and plastic limit, is a key parameter for identifying clay behaviour under varying moisture conditions. Skempton's activity index, derived from the plasticity index and clay content, has proven particularly useful for predicting piping and breaching potential in clayey soils (Resendiz 1977).

Dispersive soils typically have high plastic limits, indicating high cohesion and tensile strength when dry, but low shear resistance when wet. Soils with lower Atterberg limits tend to exhibit better engineering properties and reduced shrink–swell behaviour (Sherard *et al.* 1977). The Atterberg limits also inform the liquidity index, a measure of a soil's proximity to failure under saturated conditions.

Accurate determination of these indices is critical for designing stable foundations and managing risks in expansive and dispersive soils. In engineering practice, the Atterberg limits help define appropriate soil modification techniques, such as chemical stabilisation, to improve construction safety and durability.

3.6 REFERENCES

Baldermann A, Reinprecht V, Dietzel M (2021) Chemical weathering and progressive alteration as possible controlling factors for creeping landslides. *The Science of the Total Environment* **778**, 146300. doi:10.1016/j.scitotenv.2021.146300

Barrett-Lennard EG, Anderson CG, Holmes KW, Sinnott A (2016) High soil sodicity and alkalinity cause transient salinity in south-western Australia. *Soil Research* **54**, 407–417. doi:10.1071/SR15052

Barzegar AR, Oades JM, Rengasamy P, Giles L (1994) Effect of sodicity and salinity on disaggregation and tensile strength of an Alfisol under different cropping systems. *Soil & Tillage Research* **32**, 329–345. doi:10.1016/0167-1987(94)00421-A

Bertsch PM, Seaman JC (1999) Characterization of complex mineral assemblages: implications for contaminant transport and environmental remediation. *Proceedings of the National Academy of Sciences of the United States of America* **96**, 3350–3357. doi:10.1073/pnas.96.7.3350

Biggs AJW, Mahony KM (2004) Is soil science relevant to road infrastructure? In '13th International Soil Conservation Organisation Conference', July 2004, Brisbane, Qld. Available at https://topsoil.nserl.purdue.edu/isco/isco13/PAPERS%20A-E/BIGGS%202.pdf [verified 15 July 2025].

Boucher SC (1990) 'Field tunnel erosion: its characteristics and amelioration.' (Victorian Department of Conservation and Environment, Melbourne, Vic.)

Brink ABA, Partridge TC, Williams ABA (1984) 'Soil survey for engineers.' (Oxford University Press, New York, NY)

Chorom M, Rengasamy P (1995) Dispersion and zeta potential of pure clays as related to net particle charge under varying pH, electrolyte concentration and cation type. *European Journal of Soil Science* **46**, 657–665. doi:10.1111/j.1365-2389.1995.tb01362.x

Chorom M, Rengasamy P (1996) Effect of heating on swelling and dispersion of different cationic forms of a smectite. *Clays and Clay Minerals* **44**, 783–790. doi:10.1346/CCMN.1996.0440609

Chorom M, Rengasamy P, Murray RS (1994) Clay dispersion as influenced by pH and net particle charge of sodic soils. *Australian Journal of Soil Research* **32**, 1243–1252. doi:10.1071/SR9941243

Curtin D, Steppuhn H, Selles F (1994) Effects of magnesium on cation selectivity and structural stability of sodic soils. *Soil Science Society of America Journal* **58**, 730–737. doi:10.2136/sssaj1994.03615995005800030013x

Dexter AR (1988) Advances in characterization of soil structure. *Soil & Tillage Research* **11**, 199–238. doi:10.1016/0167-1987(88)90002-5

Emerson WW (1983) Interparticle bonding. In 'Soils: an Australian viewpoint'. pp. 477–498. (CSIRO Soils: Adelaide, SA)

Emerson WW (2002) Emerson dispersion test. In 'Soil physical measurement and interpretation for land evaluation'. (Eds N McKenzie, K Coughlan, H Cresswell) pp. 190–199. (CSIRO Publishing: Melbourne, Vic.)

Emerson WW, Bakker AC (1973) The comparative effects of exchangeable calcium, magnesium, and sodium on some physical properties of red-brown earth subsoils. II. The spontaneous dispersion of aggregates in water. *Soil Research* **11**, 151–157. doi:10.1071/SR9730151

Farahani E, Emami H, Fotovat A, Khorassani R (2019) Effect of different K:Na ratios in soil on dispersive charge, cation exchange and zeta potential. *European Journal of Soil Science* **70**, 311–320. doi:10.1111/ejss.12735

Faulkner H (2006) Piping hazard on collapsible and dispersive soils in Europe. In: 'Soil erosion in Europe'. (Eds J Boardman, J Poesen) pp. 537–562. (John Wiley & Sons, Chichester)

Goldsmith PR, Smith EH (1985) Tunnelling soils in South Auckland, New Zealand. *Engineering Geology* **22**, 1–11. doi:10.1016/0013-7952(85)90033-X

Gregory J (1989) Fundamentals of flocculation. *Critical Reviews in Environmental Control* **19**, 185–230. doi:10.1080/10643388909388365

Gupta RK, Abrol IP (1990) Reclamation and management of alkali soils. *Indian Journal of Agricultural Sciences* **60**, 1–16.

Hardie MA, Doyle RB, Cotching W, Duckett T, Zund P (2009) Dispersive soils and their management. Technical Reference Manual 36. Sustainable Land Use, Department of Primary Industries and Water, Hobart, Tas.

Hu F, Liu J, Xu C, Wang Z, Liu G, Li H, Zhao S (2018) Soil internal forces initiate aggregate breakdown and splash erosion. *Geoderma* **320**, 43–51. doi:10.1016/j.geoderma.2018.01.019

Huheey JE, Keiter EA, Keiter RL (1994) 'Inorganic chemistry.' (Harper Collins, New York, NY)

Iwata S, Tabuchi T, Warkentin BP (1995) Soil–water interactions: mechanisms and applications. (Marcel Decker: New York, NY)

Jenkins B, Morand D (2004) An overview of acid-sodic soils in two regions of New South Wales, Australia. In '3rd Australian New Zealand Soils Conference, Super Soil', 5–9 December 2004, Sydney, NSW. Available at https://www.researchgate.net/publication/253913774_An_overview_of_acid-sodic_soils_in_two_regions_of_New_South_Wales_Australia [verified 30 July 2025]

Kjellander R, Marcelja S, Quirk JP (1988) Attractive double-layer interactions between calcium clay particles. *Journal of Colloid and Interface Science* **126**, 194–211. doi:10.1016/0021-9797(88)90113-0

Laird DA (2006) Influence of layer charge on swelling of smectites. *Applied Clay Science* **34**, 74–87. doi:10.1016/j.clay.2006.01.009

Letey J (1985) Relationship between soil physical properties and crop production. In 'Advances in soil science.' Vol. 1. (Ed. BA Stewart) pp. 277–294. (Springer: New York, NY) doi:10.1007/978-1-4612-5046-38

Marchuk S, Marchuk A (2018) Effect of applied potassium concentration on clay dispersion, hydraulic conductivity, pore structure and mineralogy of two contrasting Australian soils. *Soil & Tillage Research* **182**, 35–44. doi:10.1016/j.still.2018.04.016

Marchuk A, Rengasamy P (2011) Clay behaviour in suspension is related to the ionicity of clay–cation bonds. *Applied Clay Science* **53**, 754–759. doi:10.1016/j.clay.2011.05.019

Marchuk A, Rengasamy P, McNeil A, Kumar A (2012) Nature of the clay–cation bond affects soil structure as verified by X-ray computed tomography. *Soil Research* **50**, 638–644. doi:10.1071/SR12276

McBride MB (1994) 'Environmental chemistry of soils.'(Oxford University Press: Oxford, UK)

Misono M, Ochiai E, Saito Y, Yoneda Y (1967) A new dual parameter scale for the strength of Lewis acids and bases with the evaluation of their softness. *Journal of Inorganic and Nuclear Chemistry* **29**, 2685–2691. doi:10.1016/0022-1902(67)80006-X

Norrish K (1954) The swelling of montmorillonite. *Transactions of the Faraday Society* **18**, 120–134.

Palmer B, Hazelton P (1994) Sydney Water Board St. George water area main failure analysis. In 'Soil in the city'. (Eds DR Peterson, AJ Weatherly, RE White). (The University of Melbourne: Melbourne, Vic.)

Parameswaran TG (2016) Factors controlling the dispersivity of soils and the role of zeta potential. PhD Thesis, Department of Civil Engineering, Indian Institute of Science, Bangalore, India.

Parsons DF, Bostrom M, Nostro PL, Ninham B (2011) Hofmeister effects: interplay of hydration, non-electrostatic potential and ion size. *Physical Chemistry Chemical Physics* **13**, 12352–12367. doi:10.1039/c1cp20538b

Qadir M, Schubert S (2002) Degradation processes and nutrient constraints in sodic soils. *Land Degradation & Development* **13**, 275–294. doi:10.1002/ldr.504

Qadir M, Schubert S, Oster JD, Sposito G, *et al.* (2018) High magnesium water and soils: emerging environmental and food security constraints. *The Science of the Total Environment* **642**, 1108–1117. doi:10.1016/j.scitotenv.2018.06.090

Quirk JP (1994) Interparticle forces: a basis for the interpretation of soil physical behaviour. *Advances in Agronomy* **53**, 121–183. doi:10.1016/S0065-2113(08)60614-8

Rengasamy P (2002*a*) Clay dispersion. In 'Soil physical measurement and interpretation for land evaluation'. (Eds N McKenzie, K Coughlan, H Cresswell) pp. 190–199. (CSIRO Publishing: Melbourne, Vic.)

Rengasamy P (2002*b*) Transient salinity and subsoil constraints to dryland farming in Australian sodic soils: an overview. *Australian Journal of Experimental Agriculture* **42**, 351–361. doi:10.1071/EA01111

Rengasamy P (2018) Irrigation water quality and soil structural stability: a perspective with some insights. *Agronomy (Basel)* **8**, 72. doi:10.3390/agronomy8050072

Rengasamy P, Marchuk A (2011) Cation ratio of soil structural stability (CROSS). *Australian Journal of Soil Research* **49**, 280–285. doi:10.1071/SR10105

Rengasamy P, Oades JM (1979) Interaction of monomeric and polymeric species of metal ions with clay surfaces. IV. Mixed system of Aluminium (III) and Iron (III). *Soil Research* **17**, 141–153. doi:10.1071/SR9790141

Rengasamy P, Olsson KA (1991) Sodicity and soil structure. *Soil Research* **29**, 935–952. doi:10.1071/SR9910935

Rengasamy P, Sumner ME (1998) Processes involved in sodic behaviour. In 'Sodic soils: distribution, properties, management, and environmental consequences'. (Eds ME Sumner, R Naidu) pp. 35–50. (Oxford University Press: New York, NY)

Rengasamy P, Greene RSB, Ford GW, Mehanni AH (1984) Identification of dispersive behaviour and management of red-brown earths. *Soil Research* **22**, 413–431. doi:10.1071/SR9840413

Rengasamy P, Kempers JA, Olsson KA (1991) Dispersive potential of Natrixeralfs and their crusting strength. *Clay Research* **10**, 6–10.

Rengasamy P, Olsson KA, Kirby JM (1992) Physical constraints to root growth in the subsoil. In 'Proceedings, national workshop on subsoil constraints to root growth and high soil water and nutrient use by plants'. pp. 84–91. (CSIRO Soils: Adelaide, SA)

Rengasamy P, Naidu R, Beech TA, Chan KY, Chartres C (1993) Rupture strength as related to dispersive potential in Australian soils. *Catena Supplement* **24**, 65–75.

Rengasamy P, Tavakkoli E, McDonald GK (2016) Exchangeable cations and clay dispersion: net dispersive charge, a new concept for dispersive soil. *European Journal of Soil Science* **67**, 659–665. doi:10.1111/ejss.12369

Rengasamy P, Feitosa de Lacerda C, Gheyi HR (2022) Chapter 4. Salinity, sodicity and alkalinity. In 'Subsoil constraints for crop production'. (Eds TSd Oliveira, RW Bell) pp. 83–107. (Springer Nature Switzerland: Cham)

Resendiz D (1977) Relevance of Atterberg limits in evaluating piping and breaching potential. In 'Dispersive clays, related piping and erosion in geotechnical projects'. ASTM STP 623. (Eds TL Sherard, RS Decker) pp. 341–353. (American Society for Testing and Materials: Philadelphia, PA)

Setter TL, Waters I (2003) Review of prospects for germplasm improvement for waterlogging tolerance in wheat, barley and oats. *Plant and Soil* **253**, 1–34. doi:10.1023/A:1024573305997

Shainberg I, Letey J (1984) Response of soils to sodic and saline conditions. *Hilgardia* **52**, 1–57. doi:10.3733/hilg.v52n02p057

Sherard JL, Dunnigan LP, Decker RS (1977) Some engineering problems with dispersive clays. In 'Dispersive clays, related piping and erosion in geotechnical projects'. ASTM STP 623. (Eds TL Sherard, RS Decker) pp. 3–12. (American Society for Testing and Materials: Philadelphia, PA)

Slade PG, Quirk JP, Norrish K (1991) Crystalline swelling of smectite samples in concentrated NaCl solutions in relation to layer charge. *Clays and Clay Minerals* **39**, 234–238. doi:10.1346/CCMN.1991.0390302

Smiles D, Smith C (2004) A survey of the cation content of piggery effluents and some consequences of their use to irrigate soil. *Soil Research* **42**, 231–246. doi:10.1071/SR03059

So HB, Aylmore LAG (1993) How do sodic soils behave? The effects of sodicity on soil physical behaviour. *Soil Research* **31**, 761–777. doi:10.1071/SR9930761

Sposito G (2008) 'The chemistry of soils.' 2nd edn. (Oxford University Press: New York, NY)

Sposito G (2016) 'The chemistry of soils.' 3rd edn. (Oxford University Press: New York, NY)

Sposito G, Skipper NT, Sutton R, Park S, Soper AK, Greathouse JA (1999) Surface geochemistry of the clay minerals. *Proceedings of the National Academy of Sciences of the United States of America* **96**, 3358–3364. doi:10.1073/pnas.96.7.3358

Stumpf AJ (2013) Dispersive soil hazards. In: 'Encyclopedia of natural hazards'. Encyclopedia of Earth Science Series. (Ed. PT Bobrowsky) pp. 186–188. (Springer: Dordrecht)

Sumner ME (1993) Sodic soils – new perspectives. *Soil Research* **31**, 683–750. doi:10.1071/SR9930683

Sumner ME, Miller WP, Kookana RS, Hazelton P (1998) Sodicity, dispersion and environmental quality. In 'Sodic soils: distribution, properties, management, and environmental consequences'. (Eds ME Sumner, R Naidu) pp. 149–172. (Oxford University Press: New York, NY)

Tavakkoli E, Rengasamy P, Smith E, McDonald GK (2015) The effect of cation–anion interactions on soil pH and solubility of organic carbon. *European Journal of Soil Science* **66**, 1054–1062. doi:10.1111/ejss.12294 van Olphen H (1977) 'An introduction to clay colloid chemistry.' 2nd edn. (John Wiley: New York, NY)

Wang L, Yuan X, Wang M (2020) Landslide failure mechanisms of dispersive soil slopes in seasonally frozen regions. *Advances in Civil Engineering* **2020**, 8832933. doi:10.1155/2020/8832933

Weast RC (Ed.) (1978) 'Handbook of chemistry and physics.' 58th edn. (CRC Press: Cleveland, OH)

Wilson MJ, Wilson L, Patey I (2014) The influence of individual clay minerals on formation damage of reservoir sandstones: a critical review with some new insights. *Clay Minerals* **49**, 147–164. doi:10.1180/claymin.2014.049.2.02

4

Impacts on agriculture and management of dispersive soils

4.1 INTRODUCTION

Dispersive soils represent a major challenge to sustainable agriculture, landscape stability, and environmental resilience across vast agricultural regions. These soils are defined by their tendency to disintegrate into individual clay particles when exposed to water, causing structural collapse, surface sealing, gully and tunnel erosion, and poor water infiltration. As a result, they impose multiple constraints (physical, chemical, and hydrological) that severely limit crop productivity and soil health. Beyond agricultural systems, dispersive soils also undermine the geotechnical integrity of infrastructure by reducing shear strength and increasing susceptibility to slope failure and landslides (Page *et al.* 2021).

At the core of dispersive behaviour is the interaction between the solid phase (clay minerals and exchangeable cations) and the soil solution. When the net dispersive charge (i.e. the difference between dispersive and flocculating ionic forces) exceeds a critical threshold, clay particles repel each other, resulting in structural breakdown, increased turbidity, reduced porosity, and poor aeration (Rengasamy *et al.* 2016; Liu *et al.* 2024). These changes affect root development, seedling emergence, and the movement of water and nutrients within the soil matrix. In addition, ionic imbalances in the soil solution can trigger elemental toxicities (e.g. sodium (Na^+), bicarbonate (HCO_3^-), carbonate (CO_3^{2-})) and nutrient deficiencies (e.g. calcium (Ca^{2+}), zinc (Zn^{2+}), iron (Fe^{2+})), exacerbating the constraints on plant growth (McDonald *et al.* 2020).

Globally, it is estimated that over 618 million hectares of land are affected by dispersive soils, a figure largely inferred from the classification of sodic soils (Page *et al.* 2021). However, the true extent is likely far greater, given that most soil surveys do not explicitly assess clay dispersivity. This underestimation limits the capacity to prioritise and implement remediation strategies.

Traditionally, management approaches have targeted sodicity, focusing on sodium as the primary dispersive agent. Diagnostic indices such as exchangeable sodium percentage (ESP) and sodium adsorption ratio (SAR) remain in widespread use. However, mounting evidence indicates that ESP and SAR alone do not reliably predict dispersive behaviour (Rengasamy *et al.* 2016). Instead, dispersivity is governed by a combination of factors, including clay mineralogy, ionic valence, anion composition, and electrolyte concentration (Sumner 1995; Rengasamy and Sumner 1998). This evolving understanding has led to more comprehensive metrics, such as the Cation Ratio of Soil Structural Stability (CROSS) (Rengasamy and Marchuk 2011) and the net dispersive charge (Rengasamy *et al.* 2016).

Dispersivity often co-occurs with salinity, creating saline–dispersive systems that are particularly difficult to manage. These landscapes are characterised by osmotic stress, low infiltration, and unstable pH, which together degrade soil structure and impair nutrient uptake. The interactions between salinity and dispersivity are non-linear and site specific, necessitating tailored management strategies.

This chapter explores the agricultural implications of dispersive soils and outlines a modern, multidimensional framework for their management. The following sections examine how solid phase chemistry, soil solution composition, and nutrient dynamics constrain crop growth and yield. The chapter then presents four core strategies (chemical amendments, mechanical interventions, biological approaches, and water management) designed to remediate soil structure, improve resilience, and sustain productivity. In doing so, the chapter moves beyond traditional sodicity-based models and towards a more holistic understanding of dispersive soil systems.

4.2 CONSTRAINTS TO CROP PRODUCTION IN DISPERSIVE SOILS

Crop production in dispersive soils is impeded by a complex interplay of physical and chemical limitations that affect the soil–plant system at multiple levels. These constraints originate primarily from structural degradation, ionic imbalances in the soil solution, and disruptions in nutrient dynamics. Unlike other soil limitations that can often be addressed with single agronomic interventions, dispersive soils present simultaneous, multidimensional challenges that require integrated solutions.

These constraints to crop productivity in dispersive soils can be broadly classified into three interrelated categories: (1) constraints due to solid phase chemistry; (2) constraints due to soil solution chemistry; and (3) constraints to plant nutrition.

Constraints due to solid phase chemistry refer to the fundamental mineralogical and electrochemical characteristics of the soil matrix, especially the interactions between clay particles and exchangeable cations. Unfavourable cation balances, such as high levels of sodium or magnesium relative to calcium, promote clay dispersion and aggregate breakdown. This leads to reduced macroporosity, lower permeability, impaired aeration, and greater susceptibility to crusting and compaction. These changes restrict seedling emergence, water infiltration, and root exploration, ultimately limiting a plant's ability to access critical resources.

Constraints due to soil solution chemistry encompass the influence of dissolved ions on soil physical condition, ionic balance, and pH regulation. The composition and concentration of ions, particularly sodium, bicarbonate, and carbonate species, can lead to excessive alkalinity, osmotic stress, and solubility changes in nutrients. The fluctuation of soil moisture levels intensifies these effects by driving shifts in chemical equilibria, redox reactions, and electrochemical stability. In highly alkaline or saline–dispersive environments, plant roots are exposed to combined chemical and physical stresses that compromise growth.

Constraints to plant nutrition arise when disrupted soil structure and altered chemical equilibria affect the availability, mobility, and uptake of nutrients. Structural degradation reduces water and nutrient delivery via mass flow and diffusion, while the chemical environment may cause specific nutrient toxicities or deficiencies, particularly in micronutrients like Fe, Zn, Mn, and B. In poorly aerated dispersive subsoils, redox-induced shifts can result in the accumulation of toxic reduced species or the loss of essential oxidised forms, further limiting plant performance during critical growth stages.

Importantly, these three categories do not operate in isolation. Their effects are often synergistic, where one constraint amplifies another. For instance, waterlogging due to surface dispersion leads to oxygen depletion and redox shifts, which then alter nutrient forms and reduce root metabolic activity. Similarly, co-occurring salinity and dispersivity compound physical degradation with chemical toxicity, severely limiting root growth and yield potential.

The spatial distribution of dispersive features, often concentrated in subsurface horizons, adds another layer of complexity. In dryland systems, the inability of roots to penetrate compacted dispersive subsoils limits access to stored water and nutrients, especially during terminal drought periods. Addressing these multifaceted constraints requires a holistic understanding of their interactions and the use of integrated management strategies, which are explored in the sections that follow.

4.2.1 Constraints due to solid phase chemistry

The dispersive behaviour of soils originates from inherent electrochemical instabilities within the solid phase, particularly involving clay mineralogy, surface charge characteristics, and the dominance of specific exchangeable cations. When dispersive soils are wetted, excessive repulsive forces between clay particles cause them to detach and disperse into individual colloids. This loss of structural integrity disrupts all hierarchical levels of aggregation, from microaggregates to macropores, with profound consequences for soil aeration, infiltration, strength, and biological activity.

The most immediate consequence of this structural breakdown is topsoil sealing and crusting. Dispersive topsoils often become waterlogged even after moderate rainfall due to the collapse of pore networks and the subsequent formation of impermeable surface layers. Upon drying, these soils exhibit hardsetting behaviour, characterised by increased bulk density and surface crusts that hinder seedling emergence. Meanwhile, subsoil horizons, often more dispersive, become highly compacted and exhibit low porosity, preventing root penetration and limiting the movement of air, water, and solutes.

One of the most critical limitations arising from these conditions is the restriction of root access to subsoil water, a key constraint in dryland cropping systems. Subsoil layers that exhibit dispersive and compacted properties fail to store and transmit water effectively. As a result, crops are unable to use the deeper 'green water' reserves essential for sustaining growth during periods of water deficit, especially during flowering and grain-filling stages.

Recent advances in imaging techniques have revealed the severity of these structural changes. For example, Marchuk *et al.* (2013) used non-destructive X-ray computed tomography to demonstrate that increased clay dispersion leads to substantial reductions in active porosity and saturated hydraulic conductivity. Soils with higher turbidity in dispersed suspensions, indicating greater dispersion, exhibited markedly diminished pore connectivity and water flow, regardless of whether the dominant exchangeable cation was Na^+, potassium (K^+), magnesium (Mg^{2+}), or Ca^{2+}.

A useful conceptual framework to understand this is the non-limiting water range, which defines the range of soil water content within which neither oxygen stress nor mechanical impedance limits plant growth (Wu *et al.* 2003). In dispersive soils, the non-limiting water range is exceptionally narrow. Although non-dispersive soils may support optimal aeration (>15% porosity) and manageable mechanical resistance (<1 MPa) across a wide moisture range, dispersive soils meet these conditions only in a small window. This sensitivity to drying and rewetting cycles increases the likelihood of crop failure under variable climatic conditions (Ruan *et al.* 2022; Safadoust *et al.* 2024). Furthermore, structural degradation is reinforced by ecological feedbacks. Poor root growth results in reduced organic matter inputs, weakening aggregate stability and perpetuating dispersivity. In addition, limited aeration in compacted subsoils promotes redox imbalances, which can lead to the mobilisation of phytotoxic elements such as Mn and Fe.

Together, these feedbacks maintain a persistent cycle of structural dysfunction that undermines soil health (Braidotti 2023).

The constraints imposed by solid phase chemistry in dispersive soils represent a fundamental limitation to sustainable crop production. These soils are structurally weak, hydraulically dysfunctional, and mechanically resistant, conditions that inhibit root function and reduce the efficiency of water and nutrient use. Mitigating these limitations requires a combination of chemical amendments to adjust cation balances, organic matter inputs to improve aggregation, and mechanical practices to break compacted layers and restore porosity.

4.2.2 Constraints due to soil solution chemistry

The chemistry of the soil solution plays a pivotal role in shaping the structural behaviour of dispersive soils and their influence on plant performance. While the solid phase provides a physical framework, the soil solution is a dynamic medium that mediates electrochemical interactions among clay particles, organic matter, and plant roots. In dispersive soils, the composition and ionic strength of the solution dictate the extent of clay swelling, dispersion, flocculation, and, ultimately, the soil's hydraulic conductivity and aeration capacity (Tavakkoli *et al.* 2010*b*, 2015).

4.2.2.1 Total salt concentration and electrolyte composition

The total concentration of dissolved salts in the soil solution, typically measured as electrical conductivity, has long served as an indicator of salinity-induced stress. However, in dispersive soils, the composition of electrolytes plays a more decisive role than total salinity alone. The relative abundance and activity of individual cations, especially Na^+, K^+, Ca^{2+}, and Mg^{2+}, strongly influence particle–particle interactions, with different ions exerting distinct effects on soil aggregation and structural stability (Sumner 1995).

In dryland agricultural systems, soil water content is highly variable, and fluctuations in moisture levels lead to corresponding changes in solute concentration. As soils dry, salts become concentrated, which may temporarily enhance structural stability due to increased ionic strength. However, upon rewetting, the dilution of electrolytes can reduce solution strength below the flocculation threshold, triggering a rapid collapse of soil aggregates and widespread dispersion. These transient moisture regimes introduce temporal instability in dispersive soils, rendering them particularly susceptible to degradation during wetting–drying cycles (Rengasamy and Sumner 1998).

4.2.2.2 Cation ratios and net dispersive charge

Traditional indices such as the SAR and ESP have been widely used to assess sodicity and predict dispersion risk. Nevertheless, these metrics often fail to capture the complexity of soil–cation interactions because they omit the dispersive influence of potassium and the context-dependent role of magnesium. To address these shortcomings, the CROSS index was developed by Rengasamy and Marchuk (2011), incorporating weighted contributions of Na^+, K^+, Mg^{2+}, and Ca^{2+}. Although CROSS represented a substantial advancement over SAR, it has limitations in its universality across diverse soil types and irrigation water chemistries.

A more recent and comprehensive metric, the Cationic Charge Ratio for Soil Structural Stability (CROSSc), was introduced by Rengasamy and Tavakkoli (2025) to improve predictive accuracy and conceptual clarity. This index distinguishes between dispersive charge, which is primarily conferred by monovalent cations such as Na^+ and K^+, and flocculating charge, which is associated with divalent cations such as Ca^{2+} and Mg^{2+}. The net dispersive charge, defined as the difference between the dispersive and flocculating components, serves as a mechanistic predictor of dispersion potential. When the net charge is positive, clay particles are likely to repel each other, leading to disaggregation; when it is zero or negative, the system tends towards structural stability. Unlike older indices, CROSSc reflects the physicochemical principles underpinning aggregate behaviour and has shown strong potential for guiding management decisions in both field and laboratory contexts.

4.2.2.3 Soil pH and carbonate alkalinity

Soil pH exerts a powerful influence on the structural behaviour of dispersive soils, particularly in alkaline environments where the solution chemistry becomes dominated by carbonate (CO_3^{2-}) and bicarbonate (HCO_3^-) ions (Hayward and Wadleigh 1949). At pH levels above 8.5, these anions precipitate calcium as calcium carbonate ($CaCO_3$), depleting soluble Ca^{2+}, a key flocculating ion, and simultaneously reinforcing alkalinity. This self-reinforcing cycle destabilises soil aggregates and contributes to structural degradation. The concept of carbonate alkalinity, expressed as $[HCO_3^-] + 2[CO_3^{2-}]$, is crucial for understanding both the soil's buffering capacity and its vulnerability to calcium depletion and dispersion (Van Beek and Van Breemen 1973).

Experimental work by Tavakkoli *et al.* (2015) demonstrated that the chemical form of salts applied to dispersive soils substantially affects pH dynamics and stability. Sodium chloride caused only modest pH reductions, whereas calcium chloride significantly lowered pH through carbonate precipitation, highlighting that sodium alone is not the principal driver of alkalinity. Instead, it is the imbalance between excess carbonate species and insufficient divalent cations, particularly Ca^{2+} and Mg^{2+}, that governs both alkalinity and structural decline. Therefore, effective remediation must address not just sodium replacement but also the suppression of carbonate alkalinity and restoration of ionic balance in the soil solution.

In arid and semi-arid regions, where pH commonly exceeds 9, dispersive soils experience a unique combination of abiotic stresses not captured by conventional salinity metrics. These conditions alter nutrient solubility, reduce microbial and enzymatic efficiency, and promote clay dispersion through calcium removal and enhanced cation exchange reactions. The alkalinity observed in such systems is typically linked to carbonate accumulation and reduced buffering capacity, rather than high sodium levels *per se* (McDonald *et al.* 2017; Tavakkoli *et al.* 2022*b*).

As calcium carbonate continues to precipitate, the availability of free Ca^{2+} diminishes further, weakening aggregate cohesion and allowing monovalent cations to dominate exchange sites. High pH increases the negative surface charge of clay particles, exacerbating the preferential adsorption of Na^{+} and K^{+} and amplifying dispersive behaviour. This accelerates the breakdown of aggregates, reduces permeability, and increases the risk of surface sealing, crusting, and erosion through gully and tunnel formation (Fang *et al.* 2020*a*; Sale *et al.* 2021; Uddin *et al.* 2022).

Carbonate-alkaline dispersive soils are also highly sensitive to seasonal wetting–drying cycles. In the absence of sufficient flocculating cations, even minor changes in soil moisture can trigger abrupt shifts in pH and ion equilibria. This chemical fragility is particularly problematic in landscapes irrigated with bicarbonate-rich water or where soluble salts have been leached, leaving residual carbonates behind. As pH exceeds 9.5, the risk of structural collapse intensifies, threatening crop performance, soil resilience, and long-term land productivity.

4.2.2.4 Abiotic stress from high pH and carbonate species

In many arid and semi-arid regions, dispersive soils are exposed to severe abiotic stress that extends beyond classical salinity. These soils frequently exhibit pH values of 9 or higher and are dominated by elevated concentrations of bicarbonate and carbonate ions. Such chemical conditions induce a distinctive combination of high pH stress, ionic imbalance, and mineral instability that differs fundamentally from the osmotic effects associated with neutral salts. Elevated pH reduces nutrient solubility, impairs enzymatic processes, and promotes clay dispersion through calcium precipitation and intensified cation exchange reactions. Importantly, this alkalinity is often driven not by sodicity or sodium accumulation, but by carbonate enrichment and the concurrent decline in buffering capacity (Tavakkoli *et al.* 2011, 2012; Rengasamy *et al.* 2023).

The preferential precipitation of calcium carbonate in these soils lowers the concentration of free Ca^{2+}, undermining aggregate cohesion and allowing monovalent cations, such as Na^{+} and K^{+}, to dominate the exchange complex (Ma *et al.* 2013). The increased negative charge on clay surfaces at high pH enhances the adsorption of these dispersive cations, further destabilising soil structure. As carbonate precipitation progresses, the soil's ability to buffer pH declines, accelerating aggregate breakdown, pore blockage, and reductions in both permeability and saturated hydraulic conductivity. These processes increase the risk of crusting, hardsetting, and erosion through tunnel and gully formation.

Soils with high carbonate alkalinity are also chemically unstable. As noted in the previous section, they are highly sensitive to wetting–drying cycles, particularly when irrigation involves bicarbonate-rich water or when leaching has removed soluble salts but left behind residual carbonates. Under these conditions, pH may rise above 9.5, substantially increasing the risk of structural collapse and threatening crop productivity, soil function, and long-term land sustainability (Niaz *et al.* 2023).

The stress in these systems is not merely osmotic or ionic, but reflects a more complex syndrome characterised by high pH, carbonate–cation imbalance, and severe structural vulnerability. Therefore, effective remediation therefore requires a multifaceted strategy. Although gypsum remains a valuable amendment, its success depends on addressing the full spectrum of underlying issues: reducing carbonate alkalinity, replenishing divalent cations, and restoring the soil's chemical buffering capacity. Without these targeted interventions, remediation outcomes may be short-lived, and the degradation of the soil resource will continue (Tavakkoli *et al.* 2010*a*, 2012; McDonald *et al.* 2020).

4.2.3 Constraints to plant nutrition

Plant nutrition in dispersive soils is hindered by a complex interplay of physical, chemical, and biological constraints. These limitations do not simply reflect nutrient scarcity, as in conventionally deficient soils, but rather result from environmental conditions that impair the ability of roots to access, absorb, and utilise nutrients effectively. Collectively, these constraints reduce nutrient bioavailability, disrupt uptake pathways, and suppress plant productivity, particularly under dryland and marginal conditions where dispersive soils are widespread (Rengasamy *et al.* 2023).

4.2.3.1 Physical impediments to nutrient uptake

The structural degradation associated with clay dispersion, characterised by reduced porosity, poor infiltration, and surface crusting, restricts the two principal mechanisms of nutrient transport in soil: mass flow and diffusion. In compacted or sealed dispersive soils, water movement is impeded, limiting the mobility of key nutrients such as nitrate, calcium, and magnesium. Poor soil aeration further inhibits root respiration and microbial activity, both of which are essential for nutrient mineralisation and uptake.

In dryland systems, the situation is exacerbated by transient waterlogging and drying cycles. Saturated dispersive subsoils often become temporarily anaerobic, suppressing microbial activity and altering the speciation of redox-sensitive nutrients like Fe, Mn, and S (Kyu *et al.* 2024). Upon drying, hardsetting layers inhibit root elongation, restricting the ability of crops to explore deeper or spatially variable nutrient pools. The result is a shallow, inefficient root system, unable to extract water and nutrients from beyond the uppermost soil horizons (Palta *et al.* 2010; Gangana Gowdra *et al.* 2025).

4.2.3.2 Nutrient deficiencies and toxicities in alkaline dispersive soils

Chemical constraints in dispersive soils are often driven by extreme pH conditions (Naidu and Rengasamy 1993). In acidic variants, the solubility of toxic ions such as aluminium (Al^{3+}) and manganese (Mn^{2+}) increases, whereas the availability of essential nutrients including phosphorus, calcium, and magnesium declines. In alkaline dispersive soils, particularly those with pH values above 8.5, nutrient availability is further compromised by precipitation and sorption reactions. Micronutrient deficiencies are common in these environments. Iron, for example, becomes sparingly soluble at high pH despite its abundance in total soil pools, often resulting in characteristic symptoms such as interveinal chlorosis in young leaves. Zinc availability is similarly reduced in calcareous or high-pH soils, especially in the presence of calcium carbonate. Manganese, although less soluble under oxidising conditions, may reach toxic levels in reducing, waterlogged subsoils. Boron availability also declines with increasing pH due to the sorption of borate species onto clay surfaces, although in saline–alkaline systems boron toxicity can occur where solution concentrations remain high. Phosphorus tends to form insoluble calcium–phosphate complexes at elevated pH, further limiting plant uptake and contributing to hidden hunger in crops.

In many dispersive soils, monovalent cations such as Na^+ and K^+ dominate the soil solution, whereas divalent cations like Ca^{2+} and Mg^{2+} are depleted due to precipitation or cation exchange imbalances (Vukadinović and Rengel 2007). This ionic disequilibrium not only fosters clay dispersion and structural decline but also interferes with nutrient retention and exchange on colloidal surfaces, reducing nutrient uptake efficiency.

4.2.3.3 Redox-driven nutrient imbalances

Redox dynamics further compound nutrient imbalances, particularly in dispersive subsoils that undergo periodic waterlogging. Under anaerobic conditions, the redox potential of the soil declines, shifting the valence states of several nutrients (Qadir and Schubert 2002). Iron and manganese may become more soluble and accumulate to phytotoxic levels, damaging root tissues. Aluminium, typically inactive at neutral pH, can become mobilised under highly alkaline conditions (pH >9), forming soluble aluminate species that inhibit root elongation. Sulfur may be reduced to hydrogen sulfide (H_2S), a toxic gas detrimental to both roots and soil biota. These transformations are especially detrimental during sensitive phenological stages, manifesting as chlorosis, necrosis, stunted growth, or reduced reproductive output (Fitzpatrick *et al.* 1994).

4.2.3.4 Impact on root–microbe interactions

Biological processes underpinning nutrient cycling are also adversely affected in dispersive soils (Singh 2016). Compaction and prolonged saturation suppress aerobic microbial activity, including nitrification and the establishment of mycorrhizal associations (Qadir *et al.* 2001). Elevated pH, coupled with low carbon availability in degraded dispersive soils, further reduces microbial efficiency and functional diversity. This biological impairment disrupts critical nutrient transformations and reduces the pool of plant-available nitrogen, phosphorus, and micronutrients (Wong *et al.* 2010). Moreover, the breakdown of beneficial root–microbe interactions impairs hormonal signalling, root morphology, and the rhizosphere environment, creating feedback loops that exacerbate nutrient limitations and limit crop resilience.

4.3 YIELD REDUCTION IN DISPERSIVE SOILS

Dispersive soils are a significant agronomic constraint across many agricultural landscapes, particularly in dryland cropping systems. Although the underlying physical and chemical limitations have been addressed in previous sections, the consequences of these constraints manifest agronomically as persistent yield penalties, reduced water use efficiency, diminished yield stability, and spatially variable crop performance (Braidotti 2023). These soils disrupt the synchrony between water availability, nutrient acquisition, and phenological development, leading to inefficiencies that are not easily corrected through standard management practices (Tavakkoli *et al.* 2022*b*; Uddin *et al.* 2022; Rengasamy *et al.* 2023).

Dispersive soils consistently underperform in terms of yield, even under seasons of favourable rainfall. A key agronomic challenge lies in their poor capacity to support deep and consistent root systems, which restricts access to subsoil moisture during critical reproductive phases. This is particularly detrimental in water-limited environments where yield is closely tied to effective water capture and late-season transpiration. Studies across southern Australia have demonstrated that crops grown on dispersive soils frequently experience terminal drought stress – not due to rainfall deficits *per se*, but due to their inability to utilise deeper stored water. This disconnect between rainfall and plant-available water significantly reduces crop resilience and leads to yield gaps that are often unresponsive to increased fertiliser input (Passioura and Angus 2010; Chenu *et al.* 2013).

Agronomic performance in dispersive soils is further compromised by poor water use efficiency. Crops often exhaust the upper soil moisture early in the season, whereas deeper reserves remain inaccessible due to root restriction. This decoupling of water supply from peak demand limits the crop's capacity to sustain grain filling and biomass accumulation. Kirkegaard *et al.* (2007) showed that water use during the post-anthesis period is several times more efficient in contributing to grain yield compared with early vegetative stages. Dispersive subsoils, by acting as physical and hydraulic barriers, prevent crops from capitalising on this critical efficiency window. As a result, yield potential is systematically constrained, regardless of genetic improvements or early season vigour.

The agronomic expression of dispersive soil limitations is also highly variable across space and time. Yield maps from precision agriculture platforms frequently reveal strong within-field heterogeneity in dispersive landscapes, where minor differences in clay mineralogy, sodicity, or subsoil structure result in sharp contrasts in crop response. Such variability complicates input planning and reduces the return on investment in fertiliser, seeding rates, and ameliorants. For example, areas of high subsoil constraint may exhibit stunted crops, poor nutrient uptake, and delayed maturity adjacent to more responsive zones. This micro-scale unpredictability poses a challenge for growers seeking to implement uniform agronomic strategies and underscores the need for zone-specific management, particularly for interventions such as deep ripping, gypsum incorporation, or organic amendments.

Yield quality is also affected by the agronomic implications of dispersive soils. Nutrient imbalances and water stress can lead to reduced grain protein, kernel shrivelling, and inconsistent seed size, all of which reduce market value and processing quality. Early season waterlogging, common in dispersive topsoils, may lead to poor establishment, patchy emergence, and uneven crop stands, factors that negatively affect uniformity and limit mechanised harvesting efficiency. Even when total yield is acceptable, such inconsistencies can undermine profitability and product quality, particularly in grain, legume, and oilseed systems where premium prices depend on specific quality thresholds.

From a systems perspective, dispersive soils reduce the responsiveness of crops to conventional agronomic improvements. High-input systems, including those involving advanced genetics, precision fertilisation, and conservation tillage, often fail to realise their full potential when subsoil constraints remain unaddressed. This is particularly evident in long-term trials, where the persistence of dispersive layers undermines cumulative productivity gains and increases year-to-year variability. As climate variability intensifies, these soils further exacerbate production risks, particularly in marginal rainfall zones.

Ultimately, the agronomic cost of dispersive soils is reflected not only in reduced yield but also in lost efficiency, reduced adaptability, and constrained responsiveness to innovation. Overcoming these challenges requires more than mechanical or chemical amelioration; it demands the integration of soil knowledge into agronomic decision-making, including tailored rotations, cultivar selection for root architecture, and dynamic input strategies that respond to the spatial and temporal complexity of these soils. Dispersive soils are not just chemically or physically challenging; they are also agronomically limiting, and addressing them is central to unlocking the full potential of rain-fed and irrigated cropping systems alike.

4.4 PRINCIPLES OF MANAGING DISPERSIVE SOILS

Managing dispersive soils requires a coordinated and multidimensional approach that targets both the root causes and the physical symptoms of dispersion. These soils present a uniquely complex challenge, where ionic imbalances, structural instability, and hydraulic dysfunction coexist and reinforce one another. Unlike soils constrained by isolated nutrient deficiencies or shallow compaction, dispersive soils demand holistic interventions that rebuild structure, rebalance the chemical environment, and restore biological function.

Central to dispersive behaviour is the net dispersive charge, a function of the relative dominance of monovalent cations (notably Na^+ and K^+) over divalent flocculating cations (primarily Ca^{2+} and Mg^{2+}; Rengasamy *et al.* 2016; Rengasamy and Tavakkoli 2025). In many cases, remediation begins with chemical amendments that aim to reduce sodium dominance and correct associated pH and salinity imbalances, particularly in saline–sodic and carbonate-alkaline contexts. However, although chemical strategies such as gypsum application can address ionic causes of dispersion, they are rarely sufficient in isolation. Once soil structure has collapsed, additional strategies (mechanical, biological, and hydrological) are needed to restore pore continuity, improve root proliferation, and enhance long-term functionality.

Table 4.1 presents a diagnostic matrix linking key dispersive soil conditions (e.g. pH, salinity, structural symptoms) with appropriate management principles. This framework underscores the importance of tailoring interventions to soil type, chemical condition, and landscape context.

Table 4.1 serves as a foundation for the integrated management framework described in Sections 4.5–4.8. These next sections outline the four core pillars of dispersive soil amelioration: chemical amendments, mechanical interventions, biological approaches, and water management strategies. Each pillar contributes

Table 4.1. Summary of key problems and management principles for dispersive and saline–dispersive soils across different pH conditions

This matrix highlights the combined effects of soil pH and salinity on soil physical properties and plant health, and outlines targeted remediation strategies such as flocculation, pH adjustment, and salt leaching

Soil type (pH)	Key problems	Management principles
Dispersive: acidic (pH <6)	Dispersion, swelling, hardsetting, waterlogging Acidic toxicity (Al^{3+}, Mn^{2+}), nutrient deficiency	Increase flocculating charge Apply lime to raise pH and improve Ca^{2+} availability
Dispersive: neutral (pH 6–8)	Dispersion, swelling, hardsetting, waterlogging	Increase flocculating charge via gypsum or organic Ca^{2+} sources
Dispersive: alkaline (pH >8)	Dispersion, swelling, hardsetting, waterlogging Micronutrient deficiency, carbonate alkalinity	Increase flocculating charge Reduce pH using gypsum Phytoreduction
Saline–dispersive: acidic (pH <6)	Same as dispersive soils plus: • osmotic stress from excess salts	Leach soluble salts, then Apply dispersive soil treatments (e.g. gypsum, lime)
Saline–dispersive: neutral (pH 6–8)	Same as dispersive soils plus: • osmotic stress	Leach soluble salts, then Apply dispersive soil treatments
Saline–dispersive: alkaline (pH >8)	Same as dispersive soils plus: • osmotic stress • alkaline toxicity, micronutrient deficiency	Leach soluble salts Reduce pH (if required), then Apply dispersive soil treatments

to a comprehensive remediation strategy and must be tailored to site-specific conditions, soil characteristics, and rainfall or irrigation regimes.

4.5 CHEMICAL AMENDMENTS

Chemical amendments are the cornerstone of managing dispersive soils, particularly when structural degradation is closely tied to chemical imbalances in the exchange complex and soil solution. These amendments function by displacing dispersive cations, reducing soil pH, and increasing the ionic strength of the soil solution, thereby stabilising soil aggregates and improving overall soil structure.

4.5.1 Gypsum

Gypsum ($CaSO_4 \cdot 2H_2O$) is the most widely used amendment for ameliorating dispersive soils. It supplies soluble calcium (Ca^{2+}), which replaces exchangeable sodium (Na^+) on clay surfaces, thereby reducing the net dispersive charge and promoting particle flocculation (Tavakkoli *et al.* 2022*b*). Its effectiveness is underpinned by three mechanisms:

- cation exchange: Ca^{2+} displaces Na^+ on the exchange complex, lowering ESP and improving aggregation
- an electrolyte effect: gypsum raises the ionic strength of the soil solution, suppressing dispersion even in the absence of full cation exchange
- pH moderation: in alkaline soils, gypsum reacts with sodium carbonates to form calcium carbonate, reducing soil pH and carbonate alkalinity.

Recent evidence suggests that gypsum application rates should target a minimum threshold electrolyte concentration in the soil solution, rather than relying solely on conventional 'gypsum requirement' calculations (Dang *et al.* 2018; Bennett *et al.* 2019). This ensures cost-effective treatment in rain-fed systems where water availability for gypsum dissolution is often limited.

4.5.2 Lime

In acidic dispersive soils, lime ($CaCO_3$) serves a dual purpose: neutralising soil acidity and supplying Ca^{2+}. Although it is less soluble than gypsum, lime is effective for long-term pH correction and base saturation. However, its limited solubility may reduce short-term efficacy, particularly in deep subsoil layers where immediate structural improvement is required (Bennett *et al.* 2014).

4.5.3 Organic amendments with chemical properties

Some organic materials, such as composts or manures, possess liming potential and can supply calcium and magnesium. Their ability to improve cation exchange capacity, pH buffering, and microbial activity makes them valuable in integrated amendment strategies. However, their chemical effect is generally slower than that of mineral amendments and they should be used to complement rather than replace gypsum or lime (Fang *et al.* 2020*a*, 2021, 2022).

4.5.4 Calcium chloride and other salts

Calcium chloride ($CaCl_2$) can be used in specific scenarios to provide rapidly soluble calcium and reduce dispersion. It is particularly useful for laboratory studies or small-scale interventions, but its cost typically prohibits broad-acre use (Zhang *et al.* 2023; Ziskin *et al.* 2024).

4.5.5 Emerging materials

Recent advances in material science have led to the development of new soil amendments designed to address the limitations of traditional products such as coarse gypsum and lime. Among these new soil amendments, nano-enabled materials are emerging as promising alternatives or complements, offering improved reactivity, solubility, and delivery efficiency, particularly in challenging conditions such as low-rainfall environments, subsoil-constrained profiles, and alkaline sodic landscapes where conventional amendments often underperform.

One of the most promising innovations is nano-gypsum, a finely milled form of calcium sulfate with particle sizes in the nanometre to low-micrometre range (Öksüzoğlu and Uçurum 2016; Salama *et al.* 2022). The increased surface area of nano-gypsum enhances dissolution in the soil solution, accelerating cation exchange and enabling more rapid displacement of exchangeable sodium and potassium. This leads to a quicker reduction in net dispersive charge and improved aggregate stability. The smaller particle size also improves vertical and lateral mobility in the soil profile, which is particularly beneficial in dryland systems where rainfall is episodic and amendment movement is otherwise limited (Patle and Sharma 2022). Nano-gypsum may also be suitable for site-specific or variable-rate application, given its uniform particle size and responsiveness to short wetting pulses (Salama *et al.* 2022).

However, challenges remain. Nano-gypsum is currently more expensive to produce than bulk gypsum, and its long-term environmental behaviour, such as potential nanoparticle mobility or off-target effects, requires further investigation. At this stage, it is best viewed as a precision tool for high-value or severely constrained sites rather than a broad-acre solution.

Phosphogypsum, a by-product of phosphate fertiliser manufacturing, offers a lower-cost alternative to mined gypsum with comparable calcium content and higher fineness (Gao *et al.* 2025). Its enhanced solubility makes phosphogypsum suitable for surface application and infiltration-driven movement into dispersive subsoils. However, due to potential contaminants, such as fluoride and heavy metals, its use must be regulated and quality assured (Sengupta and Dhal 2021). Where suitable sources are available, phosphogypsum could play a valuable role in large-scale soil remediation, especially in regions where natural gypsum is scarce or costly.

Further innovation is occurring in the development of engineered calcium carriers, including chelated calcium compounds, calcium-loaded nanoparticles, and encapsulated calcium formulations. These materials are designed to improve calcium solubility and targeted delivery, particularly in soils with high pH or carbonate content where traditional amendments tend to precipitate. Some are functionalised with organic ligands or biodegradable coatings to enhance reactivity, whereas others co-deliver micronutrients such as zinc or boron to simultaneously address multiple constraints. Although still largely experimental, these formulations represent a shift towards precision soil chemistry and could improve both agronomic efficiency and environmental performance in complex soil systems.

By contrast, polymeric soil conditioners such as anionic polyacrylamide (PAM) represent a more established tool for managing surface dispersion and erosion (Wallace *et al.* 1986; Wallace and Wallace 1986). PAM promotes flocculation by adhering to clay particles and forming stable aggregates through charge bridging. When applied at low concentrations, either through irrigation or as a dry granule, PAM can increase infiltration, reduce run-off, and minimise surface crusting. It is widely used in irrigated furrows and for stabilising disturbed or erosion-prone areas.

Despite these benefits, PAM has notable limitations. It degrades under ultraviolet light and microbial activity, often requiring reapplication within weeks or months. Its cost, combined with short persistence, limits its suitability for broadacre use in dryland systems. Nonetheless, PAM remains a valuable option for targeted use, such as protecting seedbeds before rainfall events, treating erodible patches, or stabilising dispersive soil on construction sites.

Natural polymers, including guar gum and other plant-derived polysaccharides, have been proposed as biodegradable alternatives to PAM (Bouranis *et al.* 1995; Ben-Hur 2006). Although promising in principle, these materials remain less studied, and their performance under variable field conditions remains inconsistent. Further research is needed to evaluate their viability, particularly for integration into organic and low-input farming systems (Panova *et al.* 2021; Lerma *et al.* 2023).

In summary, although conventional amendments like gypsum and lime continue to form the foundation of dispersive soil management, emerging materials (e.g. nano-gypsum, phosphogypsum, and engineered calcium formulations) offer exciting new possibilities. These innovations are better suited to precision agriculture, targeted remediation, and chemically complex environments. Their successful adoption will depend on continued field validation, economic feasibility, and alignment with broader agronomic and environmental objectives.

4.5.6 Placement strategies

The effectiveness of chemical amendments depends not only on product selection but also on placement relative to the dispersive zone. Surface application is appropriate for topsoil dispersion, whereas subsurface constraints require deep placement or incorporation. In rain-fed systems, where natural infiltration is limited, mechanical assistance (e.g. ripping or slotting) is often necessary to deliver amendments into the root zone and achieve structural rehabilitation at depth.

4.5.6.1 Limitations and considerations

When applying chemical amendments for dispersive soil remediation, several key limitations and site-specific considerations must be taken into account:

- leaching requirement: in saline-sodic soils, displaced Na^+ must be leached below the root zone to ensure permanent improvement
- rainfall dependency: the dissolution and movement of gypsum depend on adequate rainfall or irrigation
- targeted diagnosis: amendment selection and rate should be informed by soil testing, including measures such as net dispersive charge, ESP, pH, and CROSSc.

Chemical amendments, when matched appropriately to soil constraints and applied with consideration for placement and environmental conditions, provide the foundation for dispersive soil remediation. When integrated with physical and biological interventions, they can restore structure, improve root access to water and nutrients, and enhance long-term productivity and soil function.

4.6 MECHANICAL INTERVENTIONS

Mechanical strategies are vital components of dispersive soil amelioration, particularly in situations where structural degradation has progressed to the point that chemical amendments alone are insufficient. In these cases, interventions aim to disrupt compacted layers, restore pore connectivity, enhance root penetration, and facilitate the incorporation and movement of chemical amendments.

One of the most widely applied techniques is deep ripping, which involves fracturing compacted subsoil layers, typically beyond 30 cm depth, to reduce mechanical resistance and improve access to deeper water and nutrient stores (Riddell *et al.* 1988). However, the effectiveness of deep ripping in dispersive soils is highly contingent on soil moisture conditions, clay mineralogy, and timing. If ripping is performed when soils are excessively dry or saturated, the benefits are often short lived. Once rainfall occurs, exposed clods in dispersive subsoils may redisperse, leading to slumping and collapse of the rip lines. In some cases, this process can restore the pre-existing dense state within a single season (Back *et al.* 2024).

Further risks arise when sodic or saline subsoils are brought to the surface during ripping. In such cases, surface structure and chemistry can deteriorate, exacerbating crusting, nutrient imbalance, and water repellence. For these reasons, deep ripping is best implemented in conjunction with chemical amendments, particularly gypsum. When gypsum is applied to the surface and followed by ripping, the process enhances calcium movement into the subsoil, improving flocculation and reducing the dispersive charge at depth (Blackwell *et al.* 1991). However, in highly dispersive or low-permeability soils, vertical movement of gypsum remains slow, and mechanical mixing or slotting may be necessary to ensure even distribution of the amendment.

The use of precision ripping technologies, including adjustable-depth tines and site-specific deployment, offers new opportunities for targeted intervention in heterogeneous fields. These tools may also be coupled with liquid injection systems to deliver amendments, nutrients, or biological stimulants directly into the rip line, improving both efficiency and response in constrained zones (Ellington 1986).

Nevertheless, mechanical interventions must be carefully timed and context specific. Their success depends on adequate soil moisture, energy input, and the compatibility of soil type with the intended depth and intensity of intervention. Without a clear yield response, the costs of deep ripping, including fuel, labour, and machinery wear, may not be justified. As such, mechanical strategies should be viewed as enablers of soil rehabilitation, not stand-alone solutions. However, when used in combination with chemical amendments and biological enhancements, they can play a pivotal role in overcoming subsoil constraints and unlocking the productive potential of dispersive soils.

4.7 BIOLOGICAL APPROACHES

Biological strategies offer a long-term pathway to restoring the structure and resilience of dispersive soils. Unlike chemical or mechanical interventions, which often deliver rapid but short-lived results, biological approaches act more gradually, enhancing aggregation, porosity, microbial activity, and nutrient cycling over time. Biological approaches include the application of organic amendments, the use of deep-rooted plants, and the stimulation of rhizosphere processes to re-establish a stable, biologically active soil matrix.

The application of organic amendments, such as composts, manures, crop residues, or organic pellets, has been shown to support microbial growth and improve macroaggregation, even in sodic or alkaline subsoils. By stimulating microbial respiration and the production of polysaccharides, glomalin, and other binding agents, these materials enhance aggregate stability and infiltration. In a notable example, Clark *et al.* (2007) observed an increase in macroaggregates (>2 mm) from 1% to 42% following deep placement of wheat straw and nitrogen in a sodic clay near Ballan, Victoria, with accompanying multiyear yield benefits.

The effectiveness of organic amendments depends on their chemical composition (e.g. C:N ratio, lignin content, and the availability of labile carbon), their compatibility with soil pH and microbial communities, and adequate soil moisture to support decomposition. Recent studies suggest that co-applying gypsum with organic materials can provide additive or synergistic benefits: gypsum supplies soluble calcium and moderates pH, whereas organic materials enhance microbial function and physical structure, producing a more comprehensive and sustained improvement (Fang *et al.* 2020*a*, 2020*b*; Wang *et al.* 2020; Tavakkoli *et al.* 2022*a*; Uddin *et al.* 2022).

Primer crops, such as lucerne or phalaris, offer another biologically driven strategy for ameliorating dispersive subsoils. Their deep root systems penetrate compacted layers and create biopores that improve aeration, water infiltration, and root access for subsequent crops. However, their success in dryland systems has been variable. Structural improvements often take 4–10 years to emerge, and opportunity costs, especially in grain-only systems, can be high. In some cases, primer crops may also reduce soil water availability for the following crop due to profile drying (Nuttall *et al.* 2008). Despite these limitations, the integration of primer crops into pasture or mixed farming systems offers a viable long-term strategy for soil structural rehabilitation.

Phytoremediation involves the use of plants to chemically alter the soil environment and gradually improve structure and ionic balance. Root and microbial respiration increases CO_2 partial pressure, enhancing the dissolution of calcium carbonate and releasing Ca^{2+}. In addition, root exudation of hydrogen ions (H^+) can lower pH in the rhizosphere, mobilising native calcium, whereas root turnover contributes organic matter that supports aggregation. Some species can also reduce exchangeable Na^+ through uptake and removal in harvested biomass (Qadir *et al.* 2007). Phytoremediation is most effective in coarse- to medium-textured soils where sufficient rainfall or irrigation enables leaching of displaced salts. In arid or low-leaching environments, phytoremediation is most effective when combined with chemical or organic amendments (Srivastava 2020).

Overall, biological approaches provide a sustainable complement to physical and chemical remediation. Their success depends on site conditions, amendment quality, and integration with broader management strategies. When implemented as part of an integrated program, biological interventions can enhance structural integrity, support nutrient cycling, and promote the long-term functional recovery of dispersive soils, laying the foundation for more regenerative and resilient farming systems.

4.8 WATER MANAGEMENT

Water management is a critical pillar in the remediation of dispersive soils and the maintenance of long-term productivity. Dispersive soils are uniquely sensitive to water inputs and hydrological dynamics: excess water can rapidly trigger clay dispersion, whereas poor drainage and limited infiltration lead to waterlogging, redox stress, and restricted crop access to deeper soil moisture. Unlike more structurally stable soils, dispersive soils require carefully balanced water regimes to avoid structural collapse and to support plant growth.

4.8.1 Managing irrigation to avoid dispersion

In irrigated systems, both the quality and timing of water application are crucial. Irrigation with low-electrolyte water can dilute the soil solution below its flocculation threshold, increasing the risk of clay dispersion and structural breakdown. This is particularly problematic in sodic and carbonate-rich soils where monovalent cations, primarily Na^+ and K^+, already dominate the exchange complex.

To mitigate these risks, irrigation schedules should be closely aligned with crop demand to prevent oversaturation. Water sources should be tested for electrical conductivity, SAR, and bicarbonate concentrations. In systems with high SAR or low electrolyte water, gypsum-enriched irrigation or pretreatment of water may be necessary to maintain ionic balance. Advanced irrigation strategies, such as pulse irrigation or deficit irrigation, can help sustain a stable wetting front, minimise solubilisation losses of applied calcium, and reduce leaching of flocculating cations from the root zone.

4.8.2 Drainage and subsurface water control

Dispersive subsoils typically exhibit low hydraulic conductivity and are prone to the formation of perched water tables. Following rainfall or irrigation, transient waterlogging is common, often leading to oxygen deprivation and redox fluctuations in the root zone. These anoxic conditions promote the accumulation of phytotoxic ions such as Fe^{2+}, Mn^{2+}, and, in some cases, Al^{3+} and borate species.

To improve surface drainage and reduce ponding, landscape engineering solutions such as graded banks, spoon drains, or raised beds can be used. Controlled traffic farming also plays an important role by limiting wheel-track compaction and preserving surface infiltration capacity. In high-rainfall zones or irrigated areas with persistent subsoil constraints, deeper drainage solutions such as mole drains or tile drains may be required. These systems facilitate the removal of excess water, enhance the effectiveness of chemical amendments (e.g. gypsum), and support the leaching of displaced Na^+ beyond the rooting zone.

4.8.3 Facilitating leaching for ion removal

An essential aspect of sodic and saline–dispersive soil remediation is the removal of displaced sodium from the root zone. Effective leaching requires a reliable supply of clean water, either from rainfall or irrigation, combined with sufficient soil permeability, which may be enhanced through chemical flocculation or mechanical interventions. However, without an exit pathway, displaced ions may reaccumulate in lower horizons, undermining long-term gains.

In low-rainfall environments, leaching is often incomplete. Over time, sodium may resaturate the exchange complex during dry periods, necessitating repeated amendment applications or the use of salt-tolerant, deep-rooted plant species to cycle salts and maintain porosity. Phytoremediation, in this context, can serve as both a remediation and a maintenance tool.

4.8.4 Improving water use efficiency and crop resilience

Beyond structural remediation, improved water management in dispersive soils is key to unlocking deeper soil water stores during late-season droughts. As demonstrated by Kirkegaard *et al.* (2007), water use during reproductive stages can have two- to threefold greater yield impact than early season uptake. However, shallow root systems, common in dispersive soils, often prevent crops from accessing this critical moisture.

Management interventions that restore subsoil structure and water dynamics (through improved infiltration, storage, and drainage) can significantly reduce this yield gap. Enhancing root proliferation into deeper, treated horizons, reducing evaporative losses from surface layers, and preventing early season waterlogging all contribute to better water use efficiency and more resilient cropping systems.

Together, the four pillars of dispersive soil management (chemical, mechanical, biological, and hydrological) offer a comprehensive framework for both remediation and long-term productivity. These strategies are not mutually exclusive. Instead, they should be selected and applied in combination, tailored to specific soil constraints, rainfall patterns, and farming systems. Only through integrated management can the complex interplay of dispersive soil chemistry, structure, and water dynamics be effectively addressed.

A comparative overview of key remediation strategies and their expected outcomes is provided in Table 4.2, supporting informed decision-making for the effective management of dispersive soils.

Dispersive soils pose a significant challenge to agriculture due to their inherent instability and the complex interplay of physical and chemical constraints. Their disaggregation upon wetting, driven by imbalanced cation exchange, carbonate interactions, and low electrolyte concentrations, leads to surface sealing, poor infiltration, waterlogging, restricted rooting, and nutrient limitations. These issues are especially problematic in dryland systems where crop access to subsoil water is essential for yield stability.

This chapter has shown that dispersivity extends beyond high exchangeable sodium; it results from the combined effects of clay mineralogy, soil solution composition, and structural degradation. Diagnostic tools such as the net dispersive charge and CROSSc index now offer more precise ways to identify and manage dispersion risk.

Table 4.2. Summary of key amelioration methods for managing dispersive and sodic soils, including their expected outcomes and practical limitations

Each approach, ranging from chemical amendments like gypsum and lime to biological and physical interventions such as cover cropping and deep ripping, targets different aspects of soil structural degradation, chemical imbalance, or water management. Selection should be based on site-specific conditions, including soil pH, rainfall, and constraints to root growth or water movement. CEC, cation exchange capacity; ESP, exchangeable sodium percentage; PAM, polyacrylamide; UV, ultraviolet

Amelioration method	Expected outcomes	Limitations and considerations
Gypsum ($CaSO_4 \cdot 2H_2O$)	Supplies Ca^{2+} to displace Na^+, promoting flocculation and improving soil structure, porosity, and infiltration Reduces ESP and crusting, facilitates deeper rooting and improved yield, particularly in dry seasons	Requires sufficient moisture for dissolution; slower acting in low rainfall areas High rates may be cost-prohibitive: prioritise critical zones Ineffective on acidic soils; drainage essential for leaching Na^+ Best paired with mechanical incorporation in dense subsoils
Lime ($CaCO_3$)	Neutralises acidity in acid-sodic soils, supplies Ca^{2+}, and improves aggregation and microbial conditions in topsoil Can be used with gypsum to maintain structure in the long term	Ineffective if pH >7.0; not suitable for neutral or alkaline soils Reacts slowly (months to years) Coarse lime requires fine grinding Use only based on soil pH testing
Elemental sulfur/acids	Lowers pH in calcareous sodic soils, dissolves native $CaCO_3$ to release Ca^{2+} for flocculation Can address high pH and micronutrient deficiencies	Hazardous to handle; not practical for broadacre use Requires the presence of free lime Slow biological oxidation; moisture-dependent Typically used in small areas or on high-value crops
PAM	Promotes flocculation and reduces surface sealing, erosion, and run-off Improves infiltration and seedling emergence during heavy rainfall	Temporary effect; degrades under UV and microbial action Costly for large-scale use Requires water for activation May create slippery surfaces if over-applied Best used in high-risk zones
Deep ripping/ subsoiling	Breaks compacted subsoil, enhances aeration, root penetration, and amendment movement Improves infiltration and rooting depth when effective	Benefits may be short lived if not combined with amendments Risk of recompaction or bringing sodic subsoil to the surface Energy-intensive and costly Timing and soil moisture critical Controlled traffic farming needed after ripping
Surface water management	Slows or redirects run-off (e.g. via contour banks, raised beds) Reduces waterlogging and erosion, protects soil during high rainfall events	Requires engineering design and regular maintenance May reduce effective cropping area Structural solutions alone do not address underlying soil issues; should be used alongside soil remediation

Amelioration method	Expected outcomes	Limitations and considerations
Organic amendments (e.g. compost, manure)	Improve aggregation, microbial activity, moisture retention, and nutrient cycling Long-term enhancement of soil structure and yield potential	Bulky and variable in nutrient/salt content Transport and application can be costly High C:N materials may immobilise N; supplement with additional N if needed Best effects achieved with repeated or high-rate applications
Biochar	Increases long-term soil carbon and CEC, improves porosity and moisture retention May reduce sodicity and enhance microbial activity	High cost and large quantities required Variable outcomes depending on soil type and biochar source Needs incorporation to avoid wind loss Best used with fertilisers or compost
Cover crops/ deep-rooted vegetation	Protect soil from erosion, improve structure via root channels, and contribute organic matter Deep-rooted species aid subsoil improvement and water use	May reduce water availability for subsequent crops if not managed properly Species must tolerate initial sodic conditions Structural benefits often take multiple seasons Requires careful termination and monitoring for pests

We have outlined four integrated management strategies:

1. Chemical amendments (e.g. gypsum, lime) to restore flocculation
2. Mechanical interventions to disrupt dense layers and enable amendment movement
3. Biological approaches using organic inputs and deep-rooted species to improve aggregation and function
4. Water management to prevent dispersion, enhance drainage, and support salt leaching.

Site-specific, combined interventions, rather than single fixes, are key to restoring productivity and resilience. Moving forward, practical diagnostics, informed extension, and broader recognition of dispersive soils as distinct from salinity or sodicity will be essential to improving soil health in vulnerable landscapes.

4.9 REFERENCES

Back MP, Jefferson AJ, Ruhm CT, Blackwood CB (2024) Effects of reclamation and deep ripping on soil bulk density and hydraulic conductivity at legacy surface mines in northeast Ohio, USA. *Geoderma* **442**, 116788. doi:10.1016/j.geoderma.2024.116788

Ben-Hur M (2006) Using synthetic polymers as soil conditioners to control runoff and soil loss in arid and semi-arid regions: a review. *Soil Research* **44**, 191–204. doi:10.1071/SR05175

Bennett JM, Greene RSB, Murphy BW, Hocking P, Tongway D (2014) Influence of lime and gypsum on long-term rehabilitation of a Red Sodosol, in a semi-arid environment of New South Wales. *Soil Research* **52**, 120–128. doi:10.1071/SR13118

Bennett JM, Marchuk A, Marchuk S, Raine SR (2019) Towards predicting the soil-specific threshold electrolyte concentration of soil as a reduction in saturated hydraulic conductivity: the role of clay net negative charge. *Geoderma* **337**, 122–131. doi:10.1016/j.geoderma.2018.08.030

Blackwell P, Jayawardane N, Green T, Wood J, Blackwell J, Beatty H (1991) Subsoil macropore space of a transitional red-brown earth after either deep tillage, gypsum or both. I. Physical effects and short term changes. *Soil Research* **29**, 123–140. doi:10.1071/SR9910123

Bouranis DL, Theodoropoulos AG, Drossopoulos JB (1995) Designing synthetic polymers as soil conditioners. *Communications in Soil Science and Plant Analysis* **26**, 1455–1480. doi:10.1080/00103629509369384

Braidotti G (2023) 'Closing yield gaps on sodic soils.' (Grains Research and Development Corporation: Canberra, ACT)

Chenu K, Deihimfard R, Chapman SC (2013) Large-scale characterization of drought pattern: a continent-wide modelling approach applied to the Australian wheatbelt – spatial and temporal trends. *New Phytologist* **198**, 801–820. doi:10.1111/nph.12192

Clark GJ, Dodgshun N, Sale PWG, Tang C (2007) Changes in chemical and biological properties of a sodic clay subsoil with addition of organic amendments. *Soil Biology & Biochemistry* **39**, 2806–2817. doi:10.1016/j.soilbio.2007.06.003

Dang A, Bennett JM, Marchuk A, Marchuk S, Biggs AJW, Raine SR (2018) Validating laboratory assessment of threshold electrolyte concentration for fields irrigated with marginal quality saline-sodic water. *Agricultural Water Management* **205**, 21–29. doi:10.1016/j.agwat.2018.04.037

Ellington A (1986) Effects of deep ripping, direct drilling, gypsum and lime on soils, wheat growth and yield. *Soil & Tillage Research* **8**, 29–49. doi:10.1016/0167-1987(86)90321-1

Fang Y, Singh BP, Collins D, Armstrong R, Van Zwieten L, Tavakkoli E (2020*a*) Nutrient stoichiometry and labile carbon content of organic amendments control microbial biomass and carbon-use efficiency in a poorly structured sodic-subsoil. *Biology and Fertility of Soils* **56**, 219–233. doi:10.1007/s00374-019-01413-3

Fang Y, Singh BP, Farrell M, Van Zwieten L, Armstrong R, Chen C, Bahadori M, Tavakkoli E (2020*b*) Balanced nutrient stoichiometry of organic amendments enhances carbon priming in a poorly structured sodic subsoil. *Soil Biology & Biochemistry* **145**, 107800. doi:10.1016/j.soilbio.2020.107800

Fang Y, Singh BP, Van Zwieten L, Collins D, Pitt W, Armstrong R, Tavakkoli E (2021) Additive effects of organic and inorganic amendments can significantly improve structural stability of a sodic dispersive subsoil. *Geoderma* **404**, 115281. doi:10.1016/j.geoderma.2021.115281

Fang Y, Tavakkoli E, Weng Z, Collins D, Harvey D, Karimian N, Luo Y, Mehra P, Rose MT, Wilhelm N, Van Zwieten L (2022) Disentangling carbon stabilization in a Calcisol subsoil amended with iron oxyhydroxides: a dual-^{13}C isotope approach. *Soil Biology & Biochemistry* **170**, 108711. doi:10.1016/j.soilbio.2022.108711

Fitzpatrick R, Boucher S, Naidu R, Fritsch E (1994) Environmental consequences of soil sodicity. *Soil Research* **32**, 1069–1093. doi:10.1071/SR9941069

Gangana Gowdra VM, Lalitha BS, Halli HM, Senthamil E, Negi P, Jayadeva HM, Basavaraj PS, Harisha CB, Boraiah KM, Adavi SB, Suresha PG, Nargund R, Mohite G, Reddy KS (2025) Root growth, yield and stress tolerance of soybean to transient waterlogging under different climatic regimes. *Scientific Reports* **15**, 6968. doi:10.1038/s41598-025-91780-9

Gao L, Li R, Yang D, Bao L, Zhang N (2025) Phosphogypsum improves soil and benefits crop growth: an effective measure for utilizing solid waste resources. *Scientific Reports* **15**, 11827. doi:10.1038/s41598-025-97216-8

Hayward HE, Wadleigh CH (1949) Plant growth on saline and alkali soils. *Advances in Agronomy* **1**, 1–38. doi:10.1016/S0065-2113(08)60745-2

Kirkegaard JA, Lilley JM, Howe GN, Graham JM (2007) Impact of subsoil water use on wheat yield. *Australian Journal of Agricultural Research* **58**, 303–315. doi:10.1071/AR06285

Kyu KL, Taylor CM, Douglas CA, Malik AI, Colmer TD, Siddique KHM, Erskine W (2024) Genetic diversity and candidate genes for transient waterlogging tolerance in mungbean at the germination and seedling stages. *Frontiers in Plant Science* **15**. doi:10.3389/fpls.2024.1297096

Lerma TA, Combatt EM, Palencia M (2023) Novel multifuctional geomimetic soil conditioner based on multilayer hybrid composites of clay-PAA-lignin: synthesis and functional characterization. *European Polymer Journal* **198**, 112376. doi:10.1016/j.eurpolymj.2023.112376

Liu P, Zhu R, Zhao F, Zhao Y (2024) Enhancing dispersive soil: an experimental study on the efficacy of microbial, electrokinetics, and chemical approaches. *Sustainability* **16**, 10425. doi:10.3390/su162310425

Ma J, Wang Z-Y, Stevenson BA, Zheng X-J, Li Y (2013) An inorganic CO_2 diffusion and dissolution process explains negative CO_2 fluxes in saline/alkaline soils. *Scientific Reports* **3**, 2025. doi:10.1038/srep02025

Marchuk A, Rengasamy P, McNeill A, Kumar A (2013) Nature of the clay–cation bond affects soil structure as verified by X-ray computed tomography. *Soil Research* **50**, 683–644. doi:10.1071/SR12276

McDonald GK, Tavakkoli E, Cozzolino D, Banas K, Derrien M, Rengasamy P (2017) A survey of total and dissolved organic carbon in alkaline soils of southern Australia. *Soil Research* **55**, 617–629. doi:10.1071/SR16237

McDonald GK, Tavakkoli E, Rengasamy P (2020) Commentary: Bread wheat with high salinity and sodicity tolerance. *Frontiers in Plant Science* **11**, 1194. doi:10.3389/fpls.2020.01194

Naidu R, Rengasamy P (1993) Ion interactions and constraints to plant nutrition in Australian sodic soils. *Soil Research* **31**, 801–819. doi:10.1071/SR9930801

Niaz S, Wehr JB, Dalal RC, Kopittke PM, Menzies NW (2023) Wetting and drying cycles, organic amendments, and gypsum play a key role in structure formation and stability of sodic Vertisols. *Soil (Göttingen)* **9**, 141–154. doi:10.5194/soil-9-141-2023

Nuttall J, Davies S, Armstrong R, Peoples M (2008) Testing the primer-plant concept: wheat yields can be increased on alkaline sodic soils when an effective primer phase is used. *Australian Journal of Agricultural Research* **59**. doi:10.1071/AR07287

Öksüzoğlu B, Uçurum M (2016) An experimental study on the ultra-fine grinding of gypsum ore in a dry ball mill. *Powder Technology* **291**, 186–192. doi:10.1016/j.powtec.2015.12.027

Page KL, Dang YP, Dalal RC, Kopittke PM, Menzies NW (2021) The impact, identification and management of dispersive soils in rainfed cropping systems. *European Journal of Soil Science* **72**, 1655–1674. doi:10.1111/ejss.13070

Palta JA, Ganjeali A, Turner NC, Siddique KHM (2010) Effects of transient subsurface waterlogging on root growth, plant biomass and yield of chickpea. *Agricultural Water Management* **97**, 1469–1476. doi:10.1016/j.agwat.2010.05.001

Panova IG, Ilyasov LO, Khaidapova DD, Bashina AS, Smagin AV, Ogawa K, Adachi Y, Yaroslavov AA (2021) Soil conditioners based on anionic polymer and anionic micro-sized hydrogel: a comparative study. *Colloids and Surfaces. A, Physicochemical and Engineering Aspects* **610**, 125635. doi:10.1016/j.colsurfa.2020.125635

Passioura JB, Angus JF (2010) Chapter 2 – Improving productivity of crops in water-limited environments. In 'Advances in Agronomy'. Vol. 106. (Ed. DL Sparks) pp. 37–75. (Academic Press)

Patle T, Sharma SK (2022) Synthesis of nano-gypsum: a computational approach to encounter soil salinity and land degradation. *Computational & Theoretical Chemistry* **1217**, 113909. doi:10.1016/j.comptc.2022.113909

Qadir M, Schubert S (2002) Degradation processes and nutrient constraints in sodic soils. *Land Degradation & Development* **13**, 275–294. doi:10.1002/ldr.504

Qadir M, Schubert S, Ghafoor A, Murtaza G (2001) Amelioration strategies for sodic soils: a review. *Land Degradation & Development* **12**, 357–386. doi:10.1002/ldr.458

Qadir M, Oster JD, Schubert S, Noble AD, Sahrawat KL (2007) Phytoremediation of sodic and saline-sodic soils. *Advances in Agronomy* **96**, 197–247. doi:10.1016/S0065-2113(07)96006-X

Rengasamy P, Marchuk A (2011) Cation ratio of soil structural stability (CROSS). *Australian Journal of Soil Research* **49**, 280–285. doi:10.1071/SR10105

Rengasamy P, Sumner ME (1998) Processes involved in sodic behaviour. In 'Sodic soils: distribution, properties, management and environmental consequences'. (Eds ME Sumner, R Naidu) pp. 35–50. (Oxford University Press: New York, NY)

Rengasamy P, Tavakkoli E (2025) Cationic charge ratio (CROSSc) as an index for dispersive soils and irrigation water quality: advancing beyond SAR and CROSS. *Earth Critical Zone* **2**, 100025. doi:10.1016/j.ecz.2025.100025

Rengasamy P, Tavakkoli E, McDonald GK (2016) Exchangeable cations and clay dispersion: net dispersive charge, a new concept for dispersive soil. *European Journal of Soil Science* **67**, 659–665. doi:10.1111/ejss.12369

Rengasamy P, McDonald GK, Tavakkoli E (2023) Managing multiple constraints to crop production in dispersive soils. In 'Soil constraints and productivity'. (Eds N Bolan, MB Kirkham.) pp. 319–341. (CRC Press: Boca Raton, FL)

Riddell KM, Webster GR, Hermans JC (1988) Effects of deep ripping on chemical and physical properties of a solonetzic soil in east-central Alberta. *Soil & Tillage Research* **12**, 1–12. doi:10.1016/0167-1987(88)90051-7

Ruan R, Zhang Z, Wang Y, Guo Z, Zhou H, Tu R, Hua K, Wang D, Peng X (2022) Long-term straw rather than manure additions improved least limiting water range in a Vertisol. *Agricultural Water Management* **261**, 107356. doi:10.1016/j.agwat.2021.107356

Safadoust A, Dashtpeyma B, Mosaddeghi MR, Asgarzadeh H, Gharabaghi B (2024) Effect of water salinity and sodicity on soil least limiting water range. *Soil Science Society of America Journal* **88**, 371–386. doi:10.1002/saj2.20635

Salama AM, Abd El-Halim AE-HA, Ibrahim MM, Aiad MA, El-Shal RM (2022) Amendment with nanoparticulate gypsum enhances spinach growth in saline-sodic soil. *Journal of Soil Science and Plant Nutrition* **22**, 3377–3385. doi:10.1007/s42729-022-00893-x

Sale P, Tavakkoli E, Armstrong R, Wilhelm N, Tang C, Desbiolles J, Malcolm B, O'Leary G, Dean G, Davenport D, Henty S, Hart M (2021) Chapter six – Ameliorating dense clay subsoils to increase the yield of rain-fed crops. In 'Advances in Agronomy'. Vol. 165. (Ed. DL Sparks) pp. 249–300. (Academic Press)

Sengupta I, Dhal PK (2021) Impact of elevated phosphogypsum on soil fertility and its aerobic biotransformation through indigenous microorganisms (IMO's) based technology. *Journal of Environmental Management* **297**, 113195. doi:10.1016/j.jenvman.2021.113195

Singh K (2016) Microbial and enzyme activities of saline and sodic soils. *Land Degradation & Development* **27**, 706–718. doi:10.1002/ldr.2385

Srivastava N (2020) Reclamation of saline and sodic soil through phytoremediation. In 'Environmental concerns and sustainable development: volume 2: biodiversity, soil and waste management'. (Eds V Shukla, N Kumar) pp. 279–306. (Springer Singapore: Singapore)

Sumner ME (1995) Sodic soils: new perspectives. In 'Australian sodic soils: distribution, properties and management'. (Eds ME Naidu, P Rengasamy) pp. 1–34. (CSIRO: Melbourne, Vic.)

Tavakkoli E, Rengasamy P, Mcdonald GK (2010*a*) High concentrations of Na^+ and Cl^- ions in soil solution have simultaneous detrimental effects on growth of faba bean under salinity stress. *Journal of Experimental Botany* **61**, 4449–4459. doi:10.1093/jxb/erq251

Tavakkoli E, Rengasamy P, Mcdonald GK (2010*b*) The response of barley to salinity stress differs between hydroponics and soil systems. *Functional Plant Biology* **37**, 621–633. doi:10.1071/FP09202

Tavakkoli E, Fatehi F, Coventry S, Rengasamy P, Mcdonald GK (2011) Additive effects of Na^+ and Cl^- ions on barley growth under salinity stress. *Journal of Experimental Botany* **62**, 2189–2203. doi:10.1093/jxb/erq422

Tavakkoli E, Paull J, Rengasamy P, McDonald GK (2012) Comparing genotypic variation in faba bean (*Vicia faba* L.) in response to salinity in hydroponic and field experiments. *Field Crops Research* **127**, 99–108. doi:10.1016/j.fcr.2011.10.016

Tavakkoli E, Rengasamy P, Smith E, McDonald GK (2015) The effect of cation–anion interactions on soil pH and solubility of organic carbon. *European Journal of Soil Science* **66**, 1054–1062. doi:10.1111/ejss.12294

Tavakkoli E, Pitt W, Armstrong D, Hilderbrand S, Fang Y, Uddin S, Armstrong RD (2022*a*) Amelioration of hostile subsoils via incorporation of organic and inorganic amendments and subsequent changes in soil properties, crop water use and improved yield, in a medium rainfall zone of south-eastern Australia (GRDC: Wagga Wagga). Available at https://grdc.com.au/resources-and-publications/grdc-update-papers/tab-content/grdc-update-papers/2021/02/amelioration-of-hostile-subsoils-via-incorporation-of-organic-and-inorganic-amendments-and-subsequent-changes-in-soil-properties,-crop-water-use-and-improved-yield,-in-a-medium-rainfall-zone-of-south-eastern-australia [verified 15 July 2025].

Tavakkoli E, Uddin S, Rengasamy P, McDonald GK (2022*b*) Field applications of gypsum reduce pH and improve soil C in highly alkaline soils in southern Australia's dryland cropping region. *Soil Use and Management* **38**, 466–477. doi:10.1111/sum.12756

Uddin S, Williams SW, Aslam N, Fang Y, Parvin S, Rust J, Van Zwieten L, Armstrong R, Tavakkoli E (2022) Ameliorating alkaline dispersive subsoils with organic amendments: are productivity responses due to nutrition or improved soil structure? *Plant and Soil* **480**, 227–244. doi:10.1007/s11104-022-05569-0

Van Beek CGEM, Van Breemen N (1973) The alkalinity of alkali soils. *Journal of Soil Science* **24**, 129–136. doi:10.1111/j.1365-2389.1973.tb00748.x

Vukadinović V, Rengel Z (2007) Dynamics of sodium in saline and sodic soils. *Communications in Soil Science and Plant Analysis* **38**, 2077–2090. doi:10.1080/00103620701548811

Wallace GA, Wallace A (1986) Control of soil erosion by polymeric soil conditioners. *Soil Science* **141**, 363–367. doi:10.1097/00010694-198605000-00012

Wallace A, Wallace GA, Cha JW (1986) Mechanisms involved in soil conditioning by polymers. *Soil Science* **141**, 381–386. doi:10.1097/00010694-198605000-00017

Wang X, Sale P, Hayden H, Tang C, Clark G, Armstrong R (2020) Plant roots and deep-banded nutrient-rich amendments influence aggregation and dispersion in a dispersive clay subsoil. *Soil Biology & Biochemistry* **141**, 107664. doi:10.1016/j.soilbio.2019.107664

Wong VNL, Greene RSB, Dalal RC, Murphy BW (2010) Soil carbon dynamics in saline and sodic soils: a review. *Soil Use and Management* **26**, 2–11. doi:10.1111/j.1475-2743.2009.00251.x

Wu L, Feng G, Letey J, Ferguson L, Mitchell J, McCullough-Sanden B, Markegard G (2003) Soil management effects on the nonlimiting water range. *Geoderma* **114**, 401–414. doi:10.1016/S0016-7061(03)00052-1

Zhang L, Ge A-H, Tóth T, Yang F, Wang Z, An F (2023) Reclamation effects of distinct volumes and concentrations of CaCl2-amended brackish ice in different saline-sodic soils. *Journal of Environmental Management* **337**, 117748. doi:10.1016/j.jenvman.2023.117748

Ziskin R, Dag A, Yermiyahu U, Levy GJ (2024) Different amendments for combating soil sodicity in an olive orchard. *Agricultural Water Management* **299**, 108837. doi:10.1016/j.agwat.2024.108837

5

Irrigation and dispersive soils

5.1 INTRODUCTION

Irrigation is a cornerstone of agricultural productivity in arid and semi-arid regions, where rainfall alone is insufficient to meet crop water requirements. Yet, the use of irrigation water, especially when derived from saline, recycled, or marginal-quality sources, can introduce dissolved salts into the soil. Over time, these salts accumulate, affecting not only plant performance but also soil physical properties, including structure, porosity, and permeability. It is estimated that nearly 25% of irrigated land globally (~809 million hectares) is impacted by salinity-related issues (Mohanavelu *et al.* 2021), with implications for soil degradation, reduced water storage capacity, and contamination of groundwater resources.

Despite water covering 71% of the Earth's surface, only 2.5% constitutes fresh water, the majority of which is stored in glaciers, polar ice caps, or deep aquifers (Mishra *et al.* 2021). Consequently, access to clean irrigation water is increasingly limited. In response, farmers frequently turn to groundwater, treated municipal wastewater, or recycled industrial effluents. These sources typically contain a mixture of cations (sodium (Na^+), potassium (K^+), magnesium (Mg^{2+}), calcium (Ca^{2+})) and anions (chloride (Cl^-), sulfate (SO_4^{2-}), bicarbonate (HCO_3^-), carbonate (CO_3^{2-})), which can alter soil chemical balances and increase the dispersive charge, leading to structural instability and reduced crop productivity.

Dispersive soils, particularly those classified as sodic, are highly susceptible to structural collapse upon wetting. The dispersion of clay particles can block soil pores, reduce infiltration rates, and cause waterlogging. This chapter explores how irrigation practices contribute to the formation and intensification of dispersive soils, and critically examines the metrics and indices used to evaluate and manage irrigation water quality in such systems.

5.2 IRRIGATION LEADING TO DISPERSIVE SOILS

The increasing use of recycled wastewater and saline or brackish groundwater in irrigation has emerged as a pragmatic strategy to meet rising water demands, especially in water-scarce environments. This approach not only alleviates pressure on freshwater resources and overexploited aquifers but also contributes to food security by sustaining crop productivity (Khan *et al.* 2022). Moreover, it aligns with the principles of sustainable water management and the circular economy, wherein water is reused and essential nutrients are returned to the soil (Hagenvoort *et al.* 2019).

However, the successful and safe implementation of recycled water irrigation hinges on the application of advanced treatment technologies, robust regulatory frameworks, and ongoing monitoring. These measures are essential to mitigate environmental and health risks associated with salinity, trace elements, and pathogens (Aldughaishi *et al.* 2024).

Figure 5.1 illustrates the hydrological and geochemical processes that occur during irrigation. As water is lost through evaporation and transpiration, dissolved salts are retained in the soil profile. If not leached below the root zone or removed via drainage, these salts accumulate in the surface or subsurface horizons. The type and concentration of salts in irrigation water, particularly the balance of cations (Na^+, K^+, Mg^{2+}, Ca^{2+}) and anions (Cl^-, SO_4^{2-}, HCO_3^-, CO_3^{2-}), strongly influence the soil's chemical interactions and its tendency to disperse.

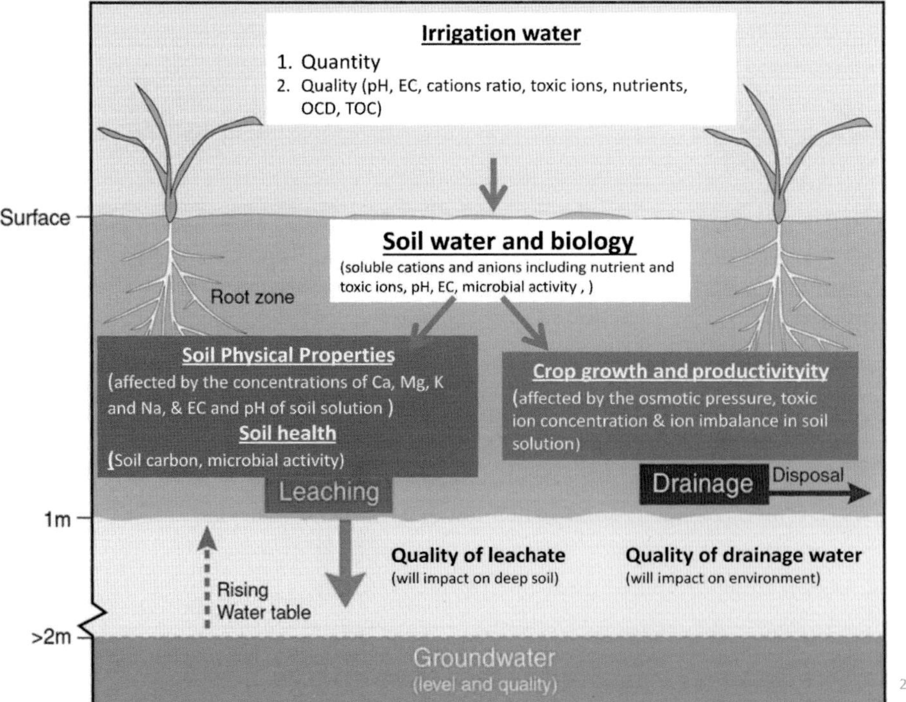

Fig. 5.1. Schematic diagram showing the processes and impacts of the quantity and quality of irrigation water.

Traditionally, sodium has been the focal point of irrigation water quality assessments due to its known detrimental effects on soil structure. High sodium levels are associated with clay dispersion, pore blockage, and reduced water infiltration. Accordingly, indices such as the exchangeable sodium percentage (ESP) and sodium adsorption ratio (SAR) have been widely adopted to predict potential soil structural decline under irrigation.

However, as discussed in earlier chapters, this sodium-centric approach is increasingly viewed as insufficient. Other cations, notably K^+ and Mg^{2+}, also influence dispersion–flocculation dynamics. Their role in modulating soil structural stability and hydraulic conductivity can be significant, particularly when irrigation water contains a mixed salt composition.

Therefore, a more holistic framework for evaluating irrigation water quality is needed, one that transcends sodium-based indices and incorporates the broader concepts of net dispersive charge, flocculating capacity, and cationic interactions. Such an approach will enable more accurate assessments of soil degradation risk and support long-term soil health and water sustainability in irrigated systems.

5.2.1 Cation Ratio of Soil Structural Stability

The Cation Ratio of Soil Structural Stability (CROSS) index was developed by Rengasamy and Marchuk (2011) as an alternative to the SAR that provides a more comprehensive assessment of irrigation water quality in relation to soil structural degradation. Unlike SAR, which considers only Na^+, Ca^{2+}, and Mg^{2+}, CROSS incorporates the relative dispersive and flocculating powers of four key cations (i.e. Na^+, K^+, Ca^{2+}, and Mg^{2+}) based on their ionicity and valence, which determine their influence on clay–cation bonding and structural stability.

The formula for CROSS is as follows:

$$CROSS = (C_{Na} + 0.56C_K)/[(C_{Ca} + 0.60C_{Mg})/2]^{0.5}$$

where C represents the molar concentrations ($mmol_c\ L^{-1}$) of each cation (Na^+, K^+, Ca^{2+}, and Mg^{2+}) in solution. The coefficient of 0.56 applied to K^+ reflects its lower dispersive power compared with Na^+, whereas the 0.60 coefficient for Mg^{2+} adjusts its lower flocculating efficiency relative to Ca^{2+}.

The CROSS index captures both the dispersive effects of Na^+ and K^+ in the numerator and the flocculating contributions of Ca^{2+} and Mg^{2+} in the denominator. As such, it is more accurate than SAR in predicting clay dispersion and related declines in hydraulic conductivity, particularly in soils with elevated levels of K^+ or Mg^{2+} (Rengasamy and Marchuk 2011; Farahani *et al.* 2019; Yan *et al.* 2024). Aldughaishi *et al.* (2024), in a study on recycled water reuse, reported that the CROSS index provided improved insight into the impact of water chemistry on soil infiltration and aggregate stability.

Despite these advantages, CROSS has been subject to critical evaluation. Reviews by Oster *et al.* (2016), Sposito (2016), and Qadir *et al.* (2021) identified several limitations, specifically:

- the relationship between CROSS and exchangeable cations has not been clearly established, unlike the well documented SAR–ESP relationship
- the predictive strength of CROSS varies across soil types, particularly with respect to clay dispersion and hydraulic conductivity
- the coefficients used for K^+ and Mg^{2+} are often soil specific and may need recalibration based on local conditions.

Nevertheless, these reviews acknowledged that CROSS generally outperforms SAR in predicting the impacts of irrigation water on soil structure and permeability. However, as with SAR, the effectiveness of CROSS is also influenced by soil-specific variables, including texture, organic matter, clay mineralogy, pH, aluminium (Al) and iron (Fe) oxides, and the precipitation of carbonates of Ca and Mg (Rengasamy 2018).

Grattan (2021), while endorsing the shift from SAR to CROSS, emphasised the need for further refinement of the CROSS model, particularly in its adaptability across different soil environments.

It is important to recognise that the assessment of irrigation water quality cannot rely on cation ratios alone. Just as salinity assessments must account for crop salt tolerance, soil structural assessments must consider the full spectrum of chemical and mineralogical interactions between irrigation water and soil. Thus, CROSS provides an important but partial framework, one that must be interpreted alongside broader soil chemistry and hydrology factors to guide sustainable irrigation management.

5.2.2 Cationic Charge Ratio for Soil Structural Stability as an index of irrigation water quality

Building on the concept of net dispersive charge, Rengasamy *et al.* (2016) highlighted that clay dispersion is driven by exchangeable cations but counteracted by the flocculating influence of cations in the soil solution. The balance between these opposing forces – dispersive and flocculating – ultimately governs the extent of clay dispersion and, in turn, the structural stability of irrigated soils. Recognising that traditional indices (such as SAR) capture only a narrow subset of these interactions, Rengasamy *et al.* (2016) advanced a more mechanistic and comprehensive framework.

In response, Rengasamy and Tavakkoli (2025) introduced the Cationic Charge Ratio for Soil Structural Stability (CROSSc) as a new index of irrigation water quality. This index integrates the relative dispersive and flocculating powers of major cations, based on their known ionicity indices and their impact on soil aggregation. It is defined as follows:

$$CROSSc = (45C_{\mathrm{Na}} + 25C_{\mathrm{K}} + 1.7C_{\mathrm{Mg}} + C_{\mathrm{Ca}})/(C_{\mathrm{Na}} + 1.8C_{\mathrm{K}} + 27C_{\mathrm{Mg}} + 45C_{\mathrm{Ca}})$$

where C denotes the concentration of cations (in $\mathrm{mmol_c\ L^{-1}}$) in either soil solution or irrigation water. The numerator represents the aggregated dispersive charge contributed by Na^+, K^+, Mg^{2+}, and Ca^{2+}, whereas the denominator represents the aggregated flocculating charge from the same cations.

CROSSc offers a direct measure of the net dispersive charge (i.e. the difference between dispersive and flocculating forces in the system) and has been shown to correlate strongly with the percentage of dispersed clay (Rengasamy *et al.* 2016). This makes it a robust indicator of the impact of irrigation water on soil structural stability, without requiring additional thresholds like the threshold electrolyte concentration.

Cation concentrations can be measured using standard analytical procedures. In practice, it is recommended that the soil be equilibrated with the intended irrigation water, and the resulting soil solution used to calculate CROSSc. This approach allows the index to reflect both the soil's buffering capacity and the chemical composition of the applied water.

Using data from 89 soils, Rengasamy and Tavakkoli (2025) demonstrated a strong linear relationship between CROSSc and dispersed clay (*Fig. 5.2*). The model captures dynamic interactions between solution-phase and exchangeable cations, and accounts for their adsorption onto soil surfaces in proportion to their ionicity and bonding strength.

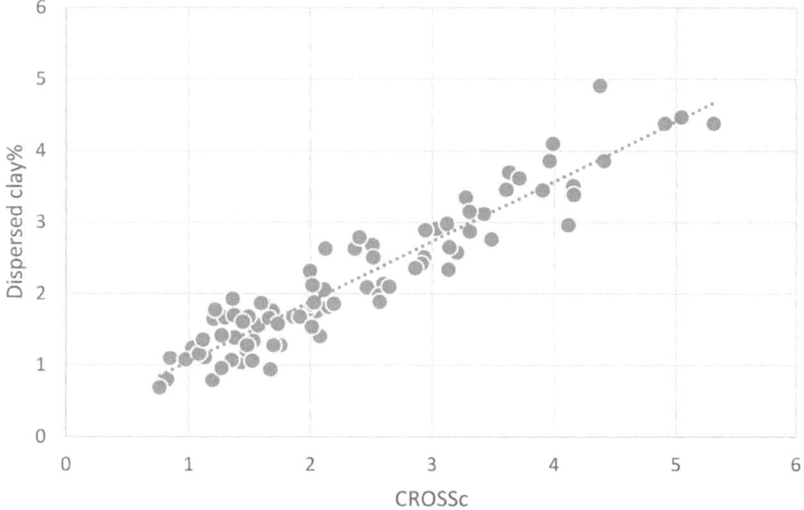

Fig. 5.2. Correlation between the Cationic Charge Ratio for Soil Structural Stability (CROSSc) and percentage of dispersed clay across 89 soil samples. Data adapted from Rengasamy and Tavakkoli (2025).

The predictive power of CROSSc lies in its ability to simultaneously incorporate the dual roles of cations in destabilising (via dispersion) and stabilising (via flocculation) soil aggregates. As the concentration of dispersive cations (e.g. Na^+, K^+) increases relative to flocculating cations (e.g. Ca^{2+}, Mg^{2+}), the net dispersive charge rises, leading to greater clay dispersion and reduced structural stability. Conversely, when flocculating cations dominate, dispersion is suppressed.

Importantly, unlike SAR or CROSS, the CROSSc model does not require separate adjustments for salinity thresholds, because the effects of ionic strength and bonding strength are inherently embedded in the ratio. Moreover, because the model is based on cation concentrations measured at the natural soil pH, it also indirectly accounts for the influence of pH on cation exchange and surface charge.

Although the initial results are promising, CROSSc, as a relatively new index, requires broader testing across different soil types, climates, and water qualities to validate its general applicability. Nonetheless, it offers a scientifically grounded, quantitatively robust framework for assessing the structural impact of irrigation water on dispersive soils, and may become a valuable tool in irrigation management and soil conservation.

5.2.3 Adsorption of dispersive cations by soil from applied irrigation water

The adsorption of cations from irrigation water into soil is governed by fundamental principles of ion exchange. One of the earliest formulations of this process was the ratio law of Schofield (1947), which described the equilibrium distribution of cations between the soil solution and the exchange complex. According to this law:

> When cations in a solution are in equilibrium with a larger number of exchangeable ions, a change in the solution concentration will not disturb equilibrium if the concentrations of all monovalent ions are altered in one ratio, all divalent ions in the square of that ratio, and all trivalent ions in the cube of that ratio (Schofield 1947).

This concept laid the foundation for the SAR, which correlates the concentration of Na^+ relative to Ca^{2+} and Mg^{2+} in solution with ESP, a widely used measure of soil sodicity and its effect on physical properties. Similar to SAR–ESP relationships, CROSS and other advanced indices can be related to the balance of monovalent and divalent ions in the exchange phase.

Marchuk *et al.* (2012) demonstrated a significant relationship between CROSS values in soil solutions and the exchangeable cation ratio (ECR%), calculated as follows:

$$ECR\% = ([\text{Exchangeable Na} + \text{K}]/\text{Effective CEC}) \times 100$$

where CEC is the cation exchange capacity. However, Marchuk *et al.* (2012) also noted that the strength of this relationship varies with CEC, underscoring the influence of soil-specific properties.

Further insights into cation behaviour come from Sposito (2016), who stated that a cation's affinity for soil surfaces increases with its tendency to form inner sphere (specific) complexes. Sposito (2016) concluded that cation affinity for soil adsorbents decreases with increasing ionicity index, as defined by the model developed by Marchuk and Rengasamy (2011; see Chapter 3). This relationship suggests the following general order of adsorption affinity under a given pH: $Ca^{2+} > Mg^{2+} > > K^{+} > Na^{+}$.

Cations forming inner sphere complexes, such as K^+ and hydrolysed forms of Al^{3+} and Fe^{3+}, are covalently bound to the surface and not readily exchangeable. These ions are typically excluded from dispersive reactions because they are structurally fixed. In contrast, outer sphere complexes, formed via electrostatic interactions, are more weakly bound and readily exchangeable; thus, they are more directly involved in dispersion and flocculation processes.

Soil pH is a critical variable in these interactions. As pH increases, the CEC also tends to rise due to increased deprotonation of functional groups. Simultaneously, pH modulates the hydrolysis of metal ions, leading to the formation of hydroxy species with varying charges and bonding behaviour. For example, alkaline irrigation waters (pH >8.5), often rich in bicarbonate and carbonate (Minhas and Qadir 2024), promote dispersion by enhancing Na^+ activity and reducing Ca^{2+} availability through carbonate precipitation. Conversely, acidic saline groundwaters (pH <5.0; Rengasamy and Olsson 1995) may induce different sorption behaviour, especially for multivalent metals like Fe^{3+} and Al^{3+}.

Although exchange isotherms are typically derived from binary systems (e.g. $Na^+ \leftrightarrow Ca^{2+}$), real-world soils are multicomponent systems. The competitive adsorption of multiple cations depends not only on their concentrations but also on their ionicity index, which reflects their relative capacity to contribute to either dispersive charge or flocculating charge.

Accordingly, the CROSSc (see Section 5.2.2) provides a more realistic representation of ion adsorption under field conditions. Because CROSSc is calculated using soil solution concentrations at natural pH, it inherently integrates the effects of both solution composition and pH on cation exchange dynamics.

As shown in Fig. 5.2, CROSSc correlates strongly with the percentage of dispersed clay and provides a direct measure of the net dispersive charge. Its strength lies in capturing the complex interactions between cation adsorption, pH-driven speciation, and soil physical behaviour, all within a single, quantifiable framework.

5.2.4 Factors affecting the accumulation of dispersive cations in soils

The accumulation of dispersive cations in soils under long-term irrigation with saline or marginal-quality water is influenced by several interconnected factors. These include the total concentration and composition of soluble salts in the applied water, soil type and mineralogy, the depth of the water table, cropping systems, climatic conditions, and the irrigation and drainage management practices used. Central to this process are two primary controls: the steady-state salt balance and the leaching fraction.

Salt and water movement within the root zone is inherently dynamic, influenced by temporal and spatial variability in inputs and outputs. The major sources of salt and water input to soils are rainfall and irrigation, whereas losses occur through evaporation, transpiration, and leaching via drainage (Minhas and Qadir 2024). The net accumulation of salts in the soil profile is determined by the balance between these inflows and outflows. Evaporation does not export salts: it merely removes water and concentrates salts in the soil. Similarly, the contribution of salt removal through plant uptake is negligible, particularly when irrigation water salinity exceeds 0.2 dS/m, because most salts are excluded at the root–soil interface. In contrast, rainwater can introduce salts, particularly in coastal regions, where salt loading from atmospheric deposition may be significant.

The leaching fraction (i.e. the proportion of irrigation water that percolates beyond the root zone) plays a vital role in regulating salt accumulation. Under steady-state conditions, the average concentration of salts in the drainage water is a function of both the leaching fraction and the salt load of the irrigation water (Minhas and Qadir 2024). A higher leaching fraction typically results in more effective salt removal. However, this model assumes that no significant precipitation or dissolution of salts occurs during the assessment period, which may not hold true in soils with carbonate or gypsum equilibria.

These dynamics have been used to estimate the leaching requirement, defined as the minimum leaching fraction necessary to maintain salt levels in the root zone below crop-specific thresholds that would otherwise reduce yield. Although this approach has traditionally focused on total salinity, as expressed by electrical conductivity, it is increasingly recognised that ionic composition, particularly the concentration of dispersive cations such as Na^+ and K^+, is equally critical in determining soil structural stability.

Over time, insufficient leaching combined with repeated irrigation using sodium-rich water can result in the progressive accumulation of dispersive cations in the soil profile. As these cations dominate the exchange complex, they increase the soil's net dispersive charge, leading to aggregate breakdown, reduced infiltration, waterlogging, and enhanced erosion risk. The risk is particularly acute in soils with low permeability or poor drainage, where salt leaching is inherently limited.

In addition, alkaline irrigation waters, often rich in bicarbonate and carbonate ions, can raise soil pH and promote sodicity through calcium and magnesium precipitation. Conversely, some saline groundwaters may have low pH and interact differently with soil colloids, further complicating management strategies (Rengasamy and Olsson 1995). Thus, both the amount and the type of salts, in addition to the pH of the irrigation water, influence the accumulation of dispersive cations and their subsequent effect on soil physical properties.

To manage these risks, a more nuanced understanding of salt dynamics is required, one that integrates chemical composition, soil adsorption processes, and hydrological behaviour. The use of mechanistic indices such as the CROSSc offers an improved basis for predicting how applied water interacts with soil and how these interactions may lead to structural degradation. When used alongside conventional salt balance and leaching calculations, CROSSc can help design irrigation strategies that minimise the build-up of dispersive cations and protect long-term soil functionality in irrigated landscapes.

5.3 IRRIGATION MANAGEMENT OF DISPERSIVE SOILS

Effective irrigation management is essential both for preventing the formation of dispersive soil layers and for mitigating the negative impacts of irrigation on already dispersive soils. This applies to irrigation with fresh water, saline groundwater, or recycled wastewater. Two major considerations must guide any management strategy: the chemistry of the irrigation water and the properties of the soil being irrigated. In particular, the composition of cations in irrigation water and the corresponding exchangeable cations in the soil determine the dispersive and flocculating charges, which influence soil structural stability (Qadir *et al.* 2018).

5.3.1 Amelioration of irrigation water

When the dispersive charge contributed by cations in irrigation water increases, it can enhance the dispersive charge of the soil's exchange complex. Therefore, modifying the chemical composition of irrigation water to reduce its dispersive charge can be an effective, and often more immediate, strategy than soil treatment. The goal is to reduce CROSSc below 1.0, ensuring that flocculating forces outweigh dispersive forces, and thereby minimising clay dispersion.

Calcium-based amendments such as gypsum ($CaSO_4{\cdot}2H_2O$) are widely used to increase the concentration of flocculating cations in irrigation water. Table 5.1 illustrates the effect of adding 15 $mmol_c\,L^{-1}$ of various chloride salts ($CaCl_2$, $MgCl_2$, KCl, and NaCl) to a representative irrigation water. The addition of $CaCl_2$ reduced CROSSc from 2.82 to 0.84, effectively reducing dispersive risk. $MgCl_2$ also decreased CROSSc (to 1.24), although less effectively than Ca^{2+}. In contrast,

Table 5.1. Changes in Cationic Charge Ratio for Soil Structural Stability, dispersive charge, and flocculating charge following the addition of 15 $mmol_c L^{-1}$ of chlorides of calcium, magnesium, potassium, and sodium to an irrigation water

CROSSc, Cationic Charge Ratio for Soil Structural Stability; DC, dispersive charge; FC, flocculating charge

Treatment	Cation composition ($mmol_c L^{-1}$)				DC ($mmol_c L^{-1}$)	FC ($mmol_c L^{-1}$)	CROSSc ($mmol_c L^{-1}$)
	Ca^{2+}	Mg^{2+}	K^+	Na^+			
Irrigation water	4.8	1.1	13.0	10.0	781.7	276.9	2.82
Addition of $CaCl_2$	19.8	1.1	13.0	10.0	796.7	951.9	0.84
Addition of $MgCl_2$	4.8	16.1	13.0	10.0	807.2	651.9	1.24
Addition of KCl	4.8	1.1	28.0	10.0	1156.7	303.9	3.81
Addition of NaCl	4.8	1.1	13.0	25.0	1456.7	291.9	4.99

adding KCl or NaCl increased CROSSc to 3.81 and 4.99 respectively, despite increasing electrolyte concentration. This reinforces the importance of cation type, not merely total salinity, in determining structural outcomes.

These findings emphasise that electrical conductivity is not an adequate indicator of irrigation water suitability when soil structural stability is at risk. Instead, CROSSc provides a more meaningful parameter by accounting for the specific roles of each cation. Notably, although potassium can contribute to dispersion, it may be fixed into the clay lattice via inner sphere complexation, which reduces its contribution to exchangeable K^+, and hence to dispersive charge (Marchuk and Marchuk 2018).

To implement these principles practically, gypsum beds have been developed to dissolve gypsum directly into irrigation water (Minhas and Qadir 2024). These systems reduce handling and application costs and can deliver up to 8 $mmol_c L^{-1}$ of dissolved Ca^{2+}. Alternatives include placing gypsum clods directly in irrigation channels (Qadir *et al.* 2007), or mechanically preparing gypsum slurries for continuous application.

Other chemical amendments, such as calcium chloride, aluminium and iron salts, pyritic materials, or sulfuric acid, can also reduce water alkalinity and dispersive charge. However, due to cost and logistical limitations, these methods are less commonly used. Emerging technologies, including electromagnetic, biological, and electrostatic treatments, are being explored to reduce salinity and dispersivity in irrigation waters, although their relevance and practicality for field-scale use remain under investigation (Rengasamy 1983).

5.3.2 Amelioration of irrigated dispersive soils

Although managing irrigation water is important, many soils already contain naturally dispersive layers, particularly within the upper 1 m of the profile (Rengasamy and Olsson 1995; Mohanavelu *et al.* 2021). These subsoil horizons typically have a very low leaching fraction (<0.02), limiting the downward movement of salts and leading to significant run-off from both rainfall and irrigation. These layers also exhibit poor water retention, thereby reducing the volume of water available to plants and necessitating more frequent irrigation to sustain crop growth.

Amelioration of these soils is a prerequisite for improving irrigation efficiency and productivity. Various inorganic and organic amendments (outlined in Chapter 4) have been used to reduce dispersive charge and restore soil structure. Among them, gypsum remains the most widely applied amendment in irrigated soils. Its application promotes the replacement of Na^+ on exchange sites with Ca^{2+}, thereby enhancing aggregation and reducing clay dispersion (Minhas and Qadir 2024).

Following gypsum application, appropriate tillage facilitates the incorporation of calcium into the soil profile, leading to structural improvements in the root zone. Over time, this also supports the accumulation of organic matter, enhanced microbial activity, and improved physical properties. As the leaching fraction increases, accumulated salts can be removed, and the CROSSc of the soil can be maintained at acceptable levels.

Where dispersivity cannot be completely eliminated, strategies such as selecting tolerant crop species (McDonald *et al.* 2020) and implementing irrigation systems tailored to site conditions can help maintain productivity. Traditional irrigation methods (border, basin, furrow) require less infrastructure but may promote surface sealing. In contrast, precision irrigation systems (sprinkler or drip) improve water distribution and reduce run-off, but require more sophisticated set-up and maintenance.

5.4 CONCLUSION

The quality of irrigation water is a critical determinant of soil structural stability and long-term productivity in irrigated systems. Irrigation waters, whether derived from surface, groundwater, or recycled sources, contain varying levels of cations such as Ca^{2+}, Mg^{2+}, K^+, and Na^+, each of which differentially affects clay dispersion. In soils with existing dispersive layers, the challenges are compounded by poor infiltration, reduced leaching, and increased waterlogging.

To maintain crop productivity, amelioration of both the irrigation water and the soil is often necessary. Gypsum remains the most practical and cost-effective amendment for reducing dispersive charge and improving flocculation. The CROSSc index, representing the ratio of dispersive to flocculating charge, provides a powerful diagnostic tool for guiding both water and soil management strategies. Keeping CROSSc below unity (<1.0) should be a central goal of irrigation management to prevent or reverse clay dispersion.

Ultimately, integrating soil chemistry, water quality, and agronomic management will be essential to sustaining productive and resilient irrigated agriculture on dispersive soils.

5.5 REFERENCES

Aldughaishi M, Grattan SR, Nicoles F, *et al.* (2024) Assessing the impact of recycled water reuse on infiltration and soil structure. *Geoderma* **452**, 117103. doi:10.1016/j.geoderma.2024.117103

Farahani E, Emami H, Fotovat A, Khorasani R (2019) Effect of different K:Na ratios in soil on dispersive charge, cation exchange and zeta potential. *European Journal of Soil Science* **70**, 311–320. doi:10.1111/ejss.12735

Grattan SR (2021) Sodicity and the soil-plant system: current assessment. In 'First IUSS Conference on Sodic Soil Reclamation', 30 July 2021, Changchun, China. (Eds Z Wang and T Tóth) pp. 22–24. (Northeast Institute of Geography and Agroecology, Chinese Academy of Sciences)

Hagenvoort J, Ortega-Reig M, Botella S, Garcia C, deLuis A, Palau Salvador G (2019) Reusing treated wastewater from a circular economy perspective – the case of the Real Acquia de Moncende in Valencia (Spain). *Water (Basel)* **11**(9), 1830. doi:10.3390/w11091830

Khan MM, Al-Haddabi MH, Akram MT, Khan MA, Farooque AA, Siddiqi SA (2022) Assessment of non-conventional irrigation water in greenhouse cucumber (*Cucumis sativus*) production. *Sustainability* **14**, 1–13.

Marchuk S, Marchuk A (2018) Effect of potassium concentration on clay dispersion, hydraulic conductivity, pore structure and mineralogy of two contrasting Australian soils. *Soil & Tillage Research* **182**, 35–44. doi:10.1016/j.still.2018.04.016

Marchuk A, Rengasamy P (2011) Clay behaviour in suspension is related to the ionicity of clay–cation bonds. *Applied Clay Science* **53**, 754–759. doi:10.1016/j.clay.2011.05.019

Marchuk A, Rengasamy P, McNeil A, Kumar A (2012) Nature of clay–cation bond affects soil structure as verified by x-ray computed tomography. *Soil Research* **50**, 638–644. doi:10.1071/SR12276

McDonald GK, Tavakkoli E, Rengasamy P (2020) Commentary: bread wheat with high salinity and sodicity tolerance. *Frontiers in Plant Science* **11**, 1194. doi:10.3389/fpls.2020.01194

Minhas PS, Qadir M (2024) Effects of irrigation with saline, saline-sodic and alkaline waters on soils. In 'Irrigation and sustainability with saline and alkali water'. (Eds PS Minhas, M Qadir) pp. 69–110. (Springer Singapore: Singapore)

Mishra B, Kumar P, Saraswat C, Chakraborty S, Gautam A (2021) Water security in a changing environment: concept, challenges and solutions. *Water (Basel)* **13**, 490. doi:10.3390/w13040490

Mohanavelu A, Naganna SR, Al-Ansari N (2021) Irrigation induced salinity and sodicity hazards on soil and groundwater: an overview of its causes, impacts and mitigation strategies. *Agriculture* **11**, 983. doi:10.3390/agriculture11100983

Oster JD, Sposito G, Smith CJ (2016) Accounting for potassium and magnesium in irrigation water quality assessment. *California Agriculture* **70**, 71–76. doi:10.3733/ca.v070n02p71

Qadir M, Oster JD, Schubert S, Noble AD, Sahrawat KL (2007) Phyto-remediation of sodic and saline-sodic soils. *Advances in Agronomy* **96**, 197–247. doi:10.1016/S0065-2113(07)96006-X

Qadir M, Schubert S, Oster JD, Sposito G (2018) High magnesium waters and soils: emerging environmental and food security constraints. *The Science of the Total Environment* **642**, 1108–1117. doi:10.1016/j.scitotenv.2018.06.090

Qadir M, Sposito G, Smith CJ, Oster JD (2021) Reassessing irrigation water quality guidelines for sodicity hazard. *Agricultural Water Management* **255**, 107054. doi:10.1016/j.agwat.2021.107054

Rengasamy P (1983) Clay dispersion in relation to changes in the electrolyte composition of dialysed red-brown earths. *European Journal of Soil Science* **34**, 723–732. doi:10.1111/j.1365-2389.1983.tb01067.x

Rengasamy P (2018) Irrigation water quality and soil structural stability: a perspective with some new insights. *Agronomy (Basel)* **8**(5), 72. doi:10.3390/agronomy8050072

Rengasamy P, Marchuk A (2011) Cation Ratio of Soil Structural Stability (CROSS). *Soil Research* **49**, 280–285. doi:10.1071/SR10105

Rengasamy P, Olsson KA (1995) Irrigation and sodicity: an overview. In 'Australian sodic soils: distribution, properties and management'. (Eds R Naidu, ME Sumner, P Rengasamy) pp. 195–203. (CSIRO Publishing: Melbourne, Vic.)

Rengasamy P, Tavakkoli E (2025) Cationic charge ratio (CROSSc) as an index for dispersive soils and irrigation water quality: advancing beyond SAR and CROSS. *Earth Critical Zone* **2**, 100025. doi:10.1016/j.ecz.2025.100025

Rengasamy P, Tavakkoli E, McDonald GK (2016) Exchangeable cations and clay dispersion: net dispersive charge, a new concept for dispersive soil. *European Journal of Soil Science* **67**, 659–665. doi:10.1111/ejss.12369

Schofield RK (1947) A ratio law governing the equilibrium of cations in soil solutions. *Proceedings of the 11th International Congress of Pure and Applied Chemistry* **3**, 257–261.

Sposito G (2016) 'The chemistry of soils.' 3rd edn. (Oxford University Press: New York, NY)

Yan S, Zhang T, Zhang B, Liu Z, Cheng Y, Feng H (2024) Cation composition of saline water affects soil structure by altering the formation of macropores and cracks in illite soils. *Soil and Tillage Research* **239**, 106052. doi:10.1016/j.still.2024.106052

6

Impacts of dispersive soils on ecosystems, environments, and infrastructure, and possible solutions

6.1 INTRODUCTION

Dispersive soils represent a significant challenge for environmental sustainability, ecological function, and the resilience of civil infrastructure. Characterised by their propensity to break down into individual clay particles upon wetting, these soils undergo rapid structural collapse, setting in motion a suite of degradation processes. When exposed to water, the electrochemical repulsion between clay particles increases, resulting in spontaneous dispersion. This leads to the deterioration of soil structure, a reduction in infiltration, impaired drainage, and compromised aeration, ultimately causing compaction and restricting root development. Consequently, the microbial activity essential for nutrient cycling diminishes, and the broader soil food web is disrupted. Plant growth is affected not only through poor rooting conditions but also through diminished access to water and nutrients.

These changes are not confined to the terrestrial realm. The erosion of dispersive soils contributes substantial sediment loads to aquatic ecosystems, where increased turbidity reduces light penetration and alters habitat structure, affecting biodiversity. Altered hydrological patterns also destabilise catchments and reduce water quality. At a landscape scale, the combined effect of structural instability and hydrological alteration transforms vegetative communities and disrupts wildlife habitats. Furthermore, dispersive soils pose a serious threat to civil infrastructure. The loss of soil cohesion and internal stability can lead to piping, tunnel collapse, and subsidence, damaging roads, low-rise buildings, embankments, irrigation channels, landfills, and earth dams. These phenomena are particularly problematic in areas undergoing land development, road construction, or extensive irrigation. Where previous chapters addressed the implications for agriculture and plant productivity, the present chapter focuses on the broader ecological, environmental, and infrastructural consequences of dispersive soils, and considers both the mechanisms of degradation and possible avenues for mitigation.

6.2 SOIL EROSION

Soil erosion in dispersive landscapes is a dominant pathway of land degradation, primarily due to the soil's physicochemical response to water. Dispersive soils are widely distributed across multiple continents, including large tracts in Australia, the US, India, Greece, South Africa, Latin America, and South-east Asia (Umesh *et al.* 2011). Unlike typical erosive processes that depend on mechanical detachment and significant water velocity, erosion in dispersive soils is primarily driven by chemical disaggregation. This phenomenon, often described as chemical piping, differs from hydraulic piping in that it stems from a loss of internal cohesion at the clay particle level rather than from high-pressure water flow.

Fig. 6.1. Piping and initiation of sinkholes on a hillside in Tasmania. The linear pattern of collapsed soil cavities reflects subsoil erosion via concentrated water flow through dispersive or unstable soil layers, ultimately leading to sinkhole formation. (Courtesy: Dr Richard Doyle, University of Tasmania)

Two clay mineralogical processes underlie this instability: swelling and deflocculation. In the swelling state, the cohesive forces among clay particles are weakened but not destroyed, leading to volume expansion. In contrast, during deflocculation, cohesion is entirely lost, and the clay disperses into a colloidal suspension. When the latter dominates, dispersive soils become highly susceptible to erosion, forming subsurface voids and ultimately resulting in structural collapse. As rainfall or irrigation water infiltrates, ionic bonds between clay particles are disrupted, and individual clay particles become suspended in water. Even low-velocity flows are sufficient to mobilise dispersed particles, distinguishing these soils from more structurally stable counterparts (Sherard *et al.* 1976). The consequences of this dispersion manifest through multiple erosion processes, including tunnel erosion, sinkhole formation, sheet and rill erosion, and, eventually, gully development (Fig. 6.1).

6.2.1 Tunnel erosion

Tunnel erosion is one of the most destructive and insidious forms of subsurface erosion in dispersive soils. It occurs when water enters through surface cracks, animal burrows, or old root channels and begins to flow laterally through the dispersive subsoil. This process often begins following land clearing or excavation activities, which expose unstable subsoils to rainfall or run-off. The infiltrating water disperses the clays, forming a slurry that moves laterally beneath the surface (Fig. 6.2). As the tunnel enlarges with successive wetting and drying cycles, the

Fig. 6.2. Tunnel erosion in a roadside soil profile, Tasmania. Subsurface pipes formed through dispersive soil layers indicate advanced stages of soil structural collapse and lateral water movement. (Courtesy: Dr Richard Doyle, University of Tasmania)

overlying soil loses support, and roof collapse leads to the appearance of potholes or gullies (Boucher 1990). Tunnel erosion is particularly hazardous because it can progress unseen for long periods before surface collapse occurs. Once formed, tunnels expand during subsequent rainfall events, compromising catchment hydrology and directly threatening built infrastructure through subsoil instability and surface subsidence. The process is both chemical, involving dispersion, and physical, involving water transport and void enlargement, and results in irreversible changes to landscape morphology and functionality (Hardie *et al.* 2007).

6.2.2 Sinkholes

Sinkholes in dispersive soils often develop as a result of subsurface voids formed through progressive soil loss. These voids originate from mechanisms similar to tunnel erosion, wherein clay particles are suspended and removed by percolating water. Jiang (2020) identified the loss of soil mass as the central mechanism underlying sinkhole formation. When the strength of the material overlying the void is insufficient to withstand applied stresses, it collapses, forming a depression at the surface.

Fig. 6.3. Numby Numby sinkhole, located in the Northern Territory, Australia. This naturally formed sinkhole features a deep central waterbody surrounded by steep rock walls and sparse vegetation. (Photograph 'Numby Numby sinkhole' by Oldbloke49, licenced under CC BY 3.0)

Sinkholes can arise under different groundwater conditions. In the vadose zone (above the water table), disintegration is the dominant mechanism, whereas seepage erosion is prevalent in unconfined aquifers. In confined aquifers, hydraulic fissuring may trigger collapse. Jiang (2020) further differentiated between unsaturated and saturated soils, recognising the role of soil saturation in determining collapse dynamics. In dispersive soils, low particle cohesion and high dispersibility accelerate the rate of soil loss, making them particularly vulnerable to sinkhole development (Fig. 6.3). Sinkholes pose substantial engineering risks, particularly when undetected voids form beneath roads, buildings, or utility corridors.

6.2.3 Sheet erosion

Sheet erosion, although less dramatic in appearance, plays a critical role in the early stages of landscape degradation. It is initiated by raindrop impact and shallow overland flow, which gradually remove the topmost layers of soil rich in nutrients and organic matter. The process is often overlooked because of its subtlety, yet its cumulative effects result in significant loss of soil fertility. Vegetation plays a crucial role in resisting sheet erosion; bare and overgrazed soils are especially vulnerable. Early indicators of sheet erosion include water puddling after light rain, the exposure of grass roots, and soil deposition around obstructions such as fence lines. As sheet flow continues, it tends to concentrate into microchannels, giving rise to rill formation.

6.2.4 Rill erosion

Rill erosion emerges as surface run-off intensifies, carving shallow channels typically less than 30 cm deep across the landscape. These rills form in zones of concentrated flow, particularly in areas with loosened or compacted soil, such as freshly cultivated paddocks or degraded pastures. Experimental studies by Shainberg *et al.* (1992) showed that dispersive soils with high exchangeable sodium and low electrolyte concentrations are particularly prone to rilling. In such cases, rainfall containing low concentrations of dissolved ions is sufficient to trigger clay dispersion and run-off. Rill formation not only accelerates soil loss but also creates preferential pathways for subsequent erosion processes. Management practices such as the use of grassed waterways, contour banking, and mulching are essential to reduce flow concentration and limit the expansion of rills into larger channels.

6.2.5 Gully erosion

Gully erosion represents the terminal stage of this erosional continuum and is often the most visually striking. It occurs when run-off, particularly during high-intensity rainfall events, gains sufficient energy to carve deep and wide channels

Fig. 6.4. Gully erosion in a dispersive soil in Tasmania. (Courtesy: Dr Richard Doyle, University of Tasmania)

into the soil (Fig. 6.4). The process is frequently initiated at vulnerable points such as stock tracks, bare patches, or root holes, where flow energy is concentrated. In dispersive soils, gully erosion is exacerbated by the collapse of pre-existing tunnels, which rapidly transform into surface gullies as the overlying material loses integrity. Once vegetation and topsoil are removed, the bare subsoil offers little resistance to further incision. Gullies can rapidly extend both upslope and downslope, altering drainage networks and transporting large volumes of sediment to downstream systems. Although gullies may eventually stabilise under certain conditions, their formation results in the permanent reconfiguration of the landscape and a substantial loss of productive land. In highly dispersive settings, soils derived from granite or sandstone are particularly susceptible, especially when ground cover is sparse during dry seasons (Crouch 1990).

6.3 DAMAGE TO CIVIL ENGINEERING INFRASTRUCTURE

Although the phenomena associated with dispersive clays were first observed by agronomists more than a century ago, their significance in civil engineering was not fully appreciated until the 1960s. Although soil scientists and agricultural engineers had already developed a sound understanding of the behaviour of dispersive clays by the 1930s, it was not until the repeated failures of small earth dams in Australia highlighted their risk potential that engineering researchers began systematically studying dispersive clay in the context of infrastructure. Pioneering work by Aitchison and Wood (1965) and later by Sherard *et al.* (1972) underscored the inability of conventional geotechnical classification tools, such as visual soil inspection, particle size distribution, specific gravity, or Atterberg limits, to reliably identify dispersive clays. As a result, specialised diagnostic protocols were developed to assess their stability and behaviour under hydraulic conditions.

6.3.1 Piping phenomena in earth dams

Dispersive clays have been used extensively in the construction of earth embankment dams around the world. However, if these materials are not properly identified and treated, they pose a significant erosion risk during the dam's operational life. Piping failure in such structures is a critical concern. This failure mode occurs when a concentrated leak forms and emerges on the downstream face of an embankment due to seepage through cracks or preferential flow paths. Initially, erosion begins at the exit point of the leak, creating localised concentrations of hydraulic force. As erosion progresses, a tunnel-like conduit forms, gradually extending upstream towards the water source. Once the pipe connects with the reservoir, the embankment is at risk of rapid and catastrophic collapse (Richards and Reddy 2007).

In cohesionless soils or those with low plasticity, piping usually results from the detachment of particles by seepage forces. In contrast, in dispersive clays, piping is initiated through chemical deflocculation along concentrated leakage channels, often microscopic cracks, rather than seepage through a porous matrix. Once dispersion begins, erosion can propagate along the full length of the channel simultaneously, bypassing the gradual enlargement typically observed in non-dispersive materials. Failures commonly occur in areas of high crack susceptibility, such as along conduits, zones of differential compressibility, or regions that have undergone desiccation.

Sherard *et al.* (1972) identified several primary mechanisms responsible for the initiation of cracks that facilitate piping in dispersive dams. These include hydraulic fracturing during foundation preparation or grouting, desiccation cracking in arid environments, drying due to delays during fill placement, differential settlement around improperly compacted conduits or structures, foundation discontinuities, and pre-existing fractures or bedding planes in bedrock that were not properly treated. Each of these factors can compromise the integrity of an embankment and act as a pathway for internal erosion. Complementing these findings, Premkumar *et al.* (2016) demonstrated that even small proportions of dispersive clays in pavement embankments can lead to contact erosion failure, particularly under fluctuating groundwater regimes.

6.3.2 Damage to civil engineering infrastructure

Civil engineering failures related to dispersive soils are governed by principles of applied mechanics and geotechnical interactions. The mechanics of solids relates the externally applied stress to internal strain responses, such as deformation, settlement, or volume change, whereas fluid mechanics accounts for hydrostatic pressures, seepage forces, and permeability characteristics (Whitlow 1995). The performance of any engineered structure is closely tied to the mechanical compatibility between the superstructure and the subsurface soil or rock foundation.

Dispersive soils, due to their structural instability, are prone to tunnel erosion, cracking, and differential settlement, all of which may occur suddenly and without significant early warning. These failures can compromise the integrity of roads, pipelines, retaining walls, and shallow foundations, resulting in costly repairs and potential safety hazards. Understanding the geological context is therefore essential. Geological analysis not only informs the availability and suitability of construction materials but also identifies potential hazards related to erosion, soil transport, and deposition. In large-scale projects such as dams, roadways, and tunnels, site-specific geological investigations are critical to identifying zones of instability and guiding design decisions to mitigate risk.

Expansive clays, including those dominated by illite, kaolinite, and montmorillonite, are a particular concern in dispersive soil contexts. Their hydrophilic nature and structural sensitivity to moisture changes contribute to swelling, shrinkage, and severe volume fluctuations. When used as subgrades in roads or as foundations for lightly loaded structures such as residential buildings or utility pipelines, these soils often lead to uneven settlement, cracking, and deformation. Soil stabilisation, via mechanical compaction or chemical treatments, is essential to improve the engineering behaviour of these problematic materials (Ikeagwuani and Nwonu 2019).

Swelling pressures generated by expansive soils can reach magnitudes capable of lifting pavements and shallow foundations, resulting in significant structural damage. In many regions, including parts of the US and Australia, the cost of maintaining roads built on expansive soils exceeds the original construction costs (Dang *et al.* 2016). Dang *et al.* (2016) emphasise the need for thorough geotechnical investigations before construction, particularly to ensure adequate drainage and subgrade treatment. Failure to do so often results in premature deterioration of pavements and engineering failures.

6.3.3 Dispersive soils and landslides

Landslides involving dispersive soils are predominantly observed on slopes where the soil structure is destabilised by water. The dispersion of clay particles significantly reduces the shear strength of the soil, especially under conditions of prolonged saturation or freeze–thaw cycling. During freeze–thaw cycles, the formation of microcracks and pore expansion permits additional water infiltration, further weakening the soil matrix and predisposing it to shallow landslide initiation. These shallow landslides often serve as precursors to deeper mass movements. Following initial failure, rainfall infiltration intensifies erosion at the top of the slope, facilitating tensile cracking and progressing the failure plane downward, ultimately resulting in deep-seated slope collapse (Wang *et al.* 2020).

Structural instability in dispersive soils stems from their inherent tendency to break down upon wetting, creating loose, fractured soil masses. This significantly lowers the soil's internal cohesion and frictional resistance. Once saturation occurs, especially during heavy rainfall events, the slope's shear resistance becomes insufficient to counteract gravitational forces, initiating failure. Rainfall-induced landslides are a globally recognised hazard, with risks heightened in regions with steep slopes, cohesive soils, and erratic precipitation patterns. Batumalai *et al.* (2023) demonstrated that both soil saturation and slope gradient are critical factors in determining failure probability, because they reduce the critical shear stress threshold and increase soil detachment rates. In extreme cases, these failures can be triggered by additional drivers such as seismic activity, leading to submarine landslides that, in turn, may generate tsunamis with catastrophic impacts on coastal infrastructure and communities.

6.3.4 Quick clays

Quick clays are a subset of highly sensitive clays characterised by an abrupt loss in shear strength upon remoulding. Their behaviour is quantified by a property known as sensitivity, which is defined as the ratio of undisturbed to remoulded shear strength. In highly sensitive soils, the post-peak undrained shear strength is drastically reduced following mechanical disturbance, leading to rapid structural collapse (Mitchell and Soga 2005). The degree of sensitivity is closely linked to the soil's dispersive properties, including ion composition, flocculation status, and clay mineralogy.

Progressive failure in quick clay systems generally requires three conditions. First, the undisturbed shear strength of the clay must be locally exceeded, often by stress concentrations or dynamic loading. Second, the material must undergo sufficient deformation to reach its remoulded state. Third, the remoulded shear strength must be low enough to allow redistribution of stress to adjacent zones, triggering a domino-like sequence of failure. Quick clay landslides are often large in scale and occur with little or no warning, posing serious risks to human life and infrastructure. Their catastrophic potential necessitates careful geotechnical evaluation in regions known to contain these materials, particularly across parts of Scandinavia and North America.

6.3.5 Soil strength and bearing capacity

The ability of soil to support structural loads without experiencing shear failure is governed by its shear strength, a fundamental parameter in geotechnical engineering. Shear strength determines the ultimate bearing capacity (i.e. the maximum pressure a soil can sustain before undergoing shear failure beneath a foundation). In dispersive and expansive soils, this capacity is often compromised by moisture fluctuations, chemical dispersion, or structural degradation. These factors must be carefully evaluated through laboratory and *in situ* testing, including triaxial shear, vane shear, and plate load tests, to ensure the safe design of footings, pavements, and embankments (Whitlow 1995).

6.4 IMPACT OF DISPERSIVE SOILS ON ECOSYSTEMS

Healthy soils are foundational to the functioning of terrestrial ecosystems and the wellbeing of human societies, underpinning a wide range of ecosystem services essential for life (Dominati *et al.* 2010). The soil ecosystem is an intricate and dynamic system comprising both biotic and abiotic components that interact across spatial and temporal scales. It supports an extraordinary diversity of life, from microbial communities to mesofauna and burrowing vertebrates, and serves as the primary medium for plant growth. Through the provision of physical support, thermal regulation, nutrients, and moisture, soils enable vegetation to thrive, forming the structural and functional basis of terrestrial ecosystems.

Soil ecosystem services arise from the interaction of abiotic soil constituents, such as mineral particles, organic carbon, soil solution, and pore air, with the living biota, which ranges in scale from genes to macro-organisms. These interactions support a wide spectrum of regulatory, provisioning, and supporting services. Agricultural soils, in particular, deliver critical services, including climate regulation through carbon sequestration, water purification, nutrient cycling, and flood mitigation via enhanced infiltration. The multifunctionality of soil is further emphasised in the concept of soil security, defined by McBratney *et al.* (2014) as comprising five dimensions: capability, condition, capital, connectivity, and codification. A systematic understanding and elicitation of these dimensions not only inform decision-making but also promote the sustainable management of soils in the face of global environmental pressures.

The regulatory and buffering functions of soils are essential for maintaining environmental quality. Soils can retain, filter, and degrade contaminants based on their physicochemical properties, thereby contributing to water quality and mitigating pollutant fluxes. However, these functions are severely compromised in dispersive soils. Characterised by their tendency to disperse upon wetting, these soils exhibit reduced structural integrity, poor permeability, and low aggregate stability, traits that impede their ability to support soil biota, filter contaminants, or store carbon.

Dispersive soils are highly degraded and increasingly vulnerable to a range of global challenges, including biodiversity loss, altered biogeochemical cycling, pollution, climate variability, and unsustainable land use change (Rockström *et al.* 2009). Their instability leads to enhanced rates of erosion and run-off, resulting in downstream water quality degradation. The elevated sediment loads and associated turbidity affect aquatic ecosystems by limiting light penetration and disrupting primary productivity. In addition, dispersive soils often carry organic matter and nutrients into water bodies, contributing to eutrophication and the proliferation of algal blooms. In many regions, these processes are compounded by the transport of chemical and microbiological contaminants, increasing the risks of ecological toxicity and public health hazards.

Soil erosion from dispersive landscapes is particularly damaging to both natural ecosystems and managed landscapes such as croplands, forests, and rangelands. The physical loss of topsoil diminishes productivity and reduces habitat quality for soil organisms. Furthermore, dispersive soils are notoriously difficult to revegetate. Surface sealing, which inhibits seed germination and root penetration, and the formation of hard, compacted layers upon drying prevent the establishment of plant cover, exacerbating erosion and further degrading the ecological function of the landscape.

Soils provide vital services such as supporting plant growth, regulating the availability and movement of water, recycling nutrients and organic matter, providing habitat for below- and above-ground organisms, moderating atmospheric composition, and serving as a foundation for human infrastructure. The capacity of soil to deliver these services is intimately linked to its physical properties, namely texture, structure, and moisture content, which, together, create the environmental matrix in which soil biota operate. Dispersive soils compromise these foundational characteristics, leading to cascading effects on biodiversity, ecosystem stability, and productivity.

In conclusion, the soil ecosystem is a vital and dynamic interface between the geosphere and biosphere, sustaining both terrestrial life and societal needs. The presence of dispersive soils, with their inherent structural weaknesses and environmental vulnerabilities, poses a significant barrier to achieving sustainable land management and ecological conservation. Understanding the interactions between the abiotic degradation processes of dispersive soils and the biological systems they support is essential for restoring functionality, promoting soil health, and safeguarding ecosystem services for future generations.

6.5 AMELIORATION OF SOIL DISPERSIVITY FOR ENGINEERING PURPOSES

The mitigation of dispersivity in clay soils is a critical engineering challenge, particularly in regions where infrastructure development intersects with problematic soils. It is now well established that soil dispersivity can be substantially reduced through targeted remediation strategies. The selection of an appropriate method depends on the nature and scale of the engineering project, budget constraints, and site-specific soil characteristics. A range of amelioration techniques has been developed, including chemical stabilisation, mechanical reinforcement, electrokinetic modification, subsurface drainage design, replacement of dispersive layers, and the use of pile foundations. In addition, the application of surface liners composed of chemically stabilised materials has proven effective in limiting clay dispersion (Vakili *et al.* 2024).

Fundamentally, engineering solutions for dispersive soils aim to reduce net dispersive forces, principally through the removal or replacement of exchangeable sodium ions and the enhancement of inter-particle flocculation. Chemical stabilisation remains the most widely adopted technique in engineering practice. Conventional additives such as lime, cement, gypsum, and magnesium oxide initiate a suite of physicochemical reactions in the soil matrix. These include cation exchange, pozzolanic hardening, and electrostatic adsorption, which, collectively, improve aggregate stability and reduce dispersion potential (Liu *et al.* 2024). For instance, calcium and magnesium ions from stabilising agents replace sodium in the exchange complex, leading to improved flocculation and reduced clay mobility.

Despite their effectiveness, traditional stabilisers pose environmental challenges. High carbon emissions, groundwater contamination risks, and adverse effects on surrounding vegetation have raised concerns regarding the sustainability of cement- and lime-based amendments. As such, increasing attention has turned towards the development of environmentally friendly alternatives. The reuse of industrial by-products, such as ground granulated blast furnace slag, waste coal gangue, and steel slag, has demonstrated potential not only in mitigating soil dispersivity but also in addressing broader issues of waste management and resource efficiency (Liu *et al.* 2024).

Recent innovations have explored the use of nanomaterials and polymers in dispersive soil treatment. Nanoparticles, through their surface van der Waals and electrostatic forces, enhance the aggregation of clay particles, resulting in reduced swelling and improved mechanical properties (Abbasi *et al.* 2018). Similarly, polyaluminum chloride has been shown to significantly decrease dispersibility while enhancing strength and resistance to erosion (Xu *et al.* 2024). Biogeotechnical approaches such as microbial-induced calcite precipitation are also emerging as promising alternatives. In microbial-induced calcite precipitation, microbial consortia induce the precipitation of calcium carbonate within the soil pore spaces, thereby cementing particles and reducing permeability and dispersivity. Li *et al.* (2021) demonstrated that this biologically driven process is mediated by pH shifts and reduced sodium exchange, both of which contribute to increased erosion resistance.

Hybrid approaches that combine multiple amelioration mechanisms have also shown success. For example, Vakili *et al.* (2018) applied a combination of lignosulfonate and electroosmosis to dispersive clay, achieving a reduction in dispersion from 88.9% to 38% with only 0.5% lignosulfonate. Polyacrylamide-based polymers, including hydrolysed polyacrylamides and copolymers of acrylamide and acrylic acid, have long been used in agricultural soil improvement, and their application is extending into geotechnical contexts due to their ability to bind particles and reduce surface sealing.

Experience in managing tunnel erosion and other forms of geotechnical failure in urban and peri-urban environments highlights the need for integrated solutions. According to Hardie *et al.* (2007), best practice in the construction of earthworks using dispersive materials involves a multipronged approach: early identification and avoidance of dispersive soils, precise recompaction, chemical treatment, sand barrier installation, topsoil burial, and revegetation. These strategies, when combined, have proven effective in stabilising vulnerable sites and extending the functional lifespan of embankments and other infrastructure. Hardie (2009) provides a comprehensive reference on practical treatments for piping, tunnelling, and rill erosion, drawing on case histories and field-scale remediation.

Importantly, amelioration strategies must be tailored to local landscape features such as slope, drainage capacity, and soil topography. Even after chemical or biological treatment, additional physical interventions, such as engineered drainage or structural reinforcement, may be required to ensure long-term stability.

Concepts underpinning these amelioration approaches are consistent with the principles of clay chemistry and electrochemical interactions discussed in Chapter 3, as well as with agricultural treatments outlined in Chapter 4 (see Table 4.2).

Central to these strategies is the manipulation of soil surface charges: effective remediation should seek to reduce the net dispersive charge below a critical threshold by decreasing the contribution from sodium and increasing the flocculating capacity through divalent or trivalent cations. This goal is best achieved through additives that promote strong covalent or hydrogen bonding to clay mineral surfaces, rather than weaker ionic associations that may be reversible under changing environmental conditions.

In conclusion, the successful amelioration of dispersive soils in engineering contexts requires an integrated understanding of soil chemistry, hydrology, and geotechnical behaviour. Advances in eco-friendly stabilisers, combined with robust physical and biological interventions, are expanding the toolkit available to engineers. A science-based, site-specific approach to dispersivity management is essential for reducing risks associated with tunnel erosion, slope failure, and foundation instability in dispersive landscapes.

6.6 REFERENCES

Abbasi N, Farjad A, Sepehri S (2018) The use of nano clay particles for stabilisation of dispersive clayey soils. *Geotechnical and Geological Engineering* **36**, 327–335. doi:10.1007/s10706-017-0330-9

Aitchison GD, Wood CC (1965) Some interactions of compaction, permeability and post construction deflocculation affecting the probability of piping failure in small earth dams. In 'Proceedings of the 6th International Conference on Soil Mechanics and Foundation Engineering, Montreal'. Vol. 2. pp. 442–446. (University of Toronto Press: Toronto)

Batumalai P, Nazer NSM, Simon N, Sulaiman N, Umar MR, Ghasali MA (2023) Soil detachment rate of a rainfall-induced landslide soil. *Water (Basel)* **15**, 2149. doi:10.3390/w15122149

Boucher SC (1990) 'Field tunnel erosion: its characteristics and amelioration.' (Monash University and Department of Conservation and Environment: Melbourne, Vic.)

Crouch RJ (1990) Erosion processes and rates for gullies in granitic soils, Bathurst, New South Wales, Australia. *Earth Surface Processes and Landforms* **15**, 169–173. doi:10.1002/esp.3290150207

Dang LC, Fatahi B, Khabbaz H (2016) Behaviour of expansive soils stabilised with hydrated lime and bagasse fibres. *Procedia Engineering* **143**, 658–665. doi:10.1016/j.proeng.2016.06.093

Dominati E, Patterson M, Mackay A (2010) A framework for classifying and quantifying the natural capital and ecosystem services of soils. *Ecological Economics* **69**, 1858–1868. doi:10.1016/j.ecolecon.2010.05.002

Hardie MA (2009) 'Dispersive soils and their management: a technical reference manual.' (Department of Primary Industries and Water: Hobart, Tas.)

Hardie MA, Cotching WE, Zund PR (2007) Rehabilitation of field tunnel erosion using techniques developed for construction with dispersive soils. *Soil Research* **45**, 280–287. doi:10.1071/SR06154

Ikeagwuani CC, Nwonu DC (2019) Emerging trends in expansive soil stabilisation: a review. *Journal of Rock Mechanics and Geotechnical Engineering* **11**, 423–440. doi:10.1016/j.jrmge.2018.08.013

Jiang F (2020) Criteria of sinkhole formation in soils from physical models. *Bulletin of Engineering Geology and the Environment* **79**, 3833–3841. doi:10.1007/s10064-020-01768-0

Li C, Shi GY, Wu HM, Wang C, Gao Y (2021) Experimental study on bio-mineralisation for dispersed soils by means of biological calcite precipitation technology. *Yantu Lixue* **42**, 333–342.

Liu P, Zhu R, Zhao F, Zhao Y (2024) Enhancing dispersive soil: an experimental study on the efficacy of microbial, electrokinetic and chemical approaches. *Sustainability* **16**, 10425. doi:10.3390/su162310425

McBratney A, Field DJ, Koch A (2014) The dimensions of soil security. *Geoderma* **213**, 203–213. doi:10.1016/j.geoderma.2013.08.013

Mitchell JK, Soga K (2005) Conduction phenomena. In 'Fundamentals of soil behaviour'. (Eds JK Mitchell, K Soga) pp. 251–320. (John Wiley & Sons, Inc.: Toronto)

Premkumar S, Piratheepan J, Arulrajah A, Disfani MM (2016) Experimental study on contact erosion failure in pavement embankment with dispersive clay. *Journal of Materials in Civil Engineering* **28**, 1–9. doi:10.1061/(ASCE)MT.1943-5533.0001452

Richards KS, Reddy KR (2007) Critical appraisal of piping phenomena in earth dams. *Bulletin of Engineering Geology and the Environment* **66**, 381–402. doi:10.1007/s10064-007-0095-0

Rockström J, Steffen W, Noone K, Persson A, *et al.* (2009) Planetary boundaries: exploring the safe operating for humanity. *Ecology and Society* **14**, 32. doi:10.5751/ES-03180-140232

Shainberg I, Warrington D, Laflen JM (1992) Soil dispersibility, rain properties and slope interactions in rill formation and erosion. *Soil Science Society of America Journal* **56**, 278–283. doi:10.2136/sssaj1992.03615995005600010044x

Sherard JL, Decker RS, Rykes NL (1972) Piping in earth dams of dispersive clays. In 'Performance on earth and earth-supported structures'. Vol. 1. pp. 587–626. (American Society of Civil Engineers: New York, NY)

Sherard JL, Dunnigan LP, Decker RS (1976) Identification and nature of dispersive soils. *Journal of the Geotechnical Engineering Division* **102**, 287–301. doi:10.1061/AJGEB6.0000256

Umesh TS, Dinesh SV, Sivapulliah PV (2011) Characterisation of dispersive soil. *Materials Sciences and Applications* **2**, 629–633. doi:10.4236/msa.2011.26085

Vakili AH, Ghasemi J, bin Selamat MR, Salimi M, Farhadi MS (2018) Internal erosional behaviour of dispersive clay stabilised with lignosulfonate and reinforced with polypropylene fibre. *Construction & Building Materials* **193**, 405–415. doi:10.1016/j.conbuildmat.2018.10.213

Vakili AH, Salimi M, Keskin I, Jamalimoghadam M (2024) A systematic review of strategies for identifying and stabilizing dispersive clay soils for sustainable infrastructure. *Soil & Tillage Research* **239**, 106036. doi:10.1016/j.still.2024.106036

Wang L, Yan X, Wang M (2020) Landside failure mechanisms of dispersive soil slope in seasonally frozen regions. *Advances in Civil Engineering* **2020**, 8832933. doi:10.1155/2020/8832933

Whitlow R (1995) 'Basic soil mechanics.' 3rd edn. (Longman Scientific & Technical: Harlow)

Xu X, Lei H, Wang Q, Yan X, Guo L, Yu Z (2024) Polyaluminium chloride (PAC) modification of dispersive soil: a comprehensive study on dispersivity, mechanical properties, and micro scale mechanisms. *Construction & Building Materials* **425**, 135890. doi:10.1016/j.conbuildmat.2024.135890

7

Identification of dispersive soils

7.1. INTRODUCTION

Dispersive soils are structurally unstable clay-rich soils that readily disintegrate into individual particles upon wetting, particularly under low-electrolyte conditions such as rainfall. This phenomenon is primarily caused by a dominance of sodium ions relative to calcium and magnesium on the soil's cation exchange complex, a condition referred to as sodicity. Once dispersed, clay particles repel one another, breaking down soil aggregates, sealing pores, and initiating erosion processes such as crusting, tunnel erosion, and piping.

Dispersive soils are problematic across both agriculture and civil engineering. In agricultural systems, dispersive soils impede water infiltration, reduce aeration, and restrict root penetration, leading to lower crop productivity (see Chapter 4). In engineering, they pose serious risks to the structural integrity of embankments, canal linings, and dam walls (see Chapter 6).

Dispersive soils have been documented in over 75 countries (Vakili *et al.* 2024), reflecting their global prevalence. In Australia, they are widespread across agricultural and rangeland landscapes, requiring systematic identification to manage risks effectively. Surface indicators, such as cloudy water in puddles, poor vegetative growth, or visible gully erosion, may suggest the presence of dispersive subsoils. However, subsurface layers may not always be visibly identifiable, necessitating targeted laboratory or in-field diagnostics.

This chapter provides a comprehensive guide to identifying dispersive soils. It begins with traditional physical tests, including the crumb, pinhole, and double hydrometer methods, followed by chemical indices such as exchangeable sodium percentage (ESP) and sodium adsorption ratio (SAR), as well as newer metrics including the Cation Ratio of Soil Structural Stability (CROSS) and its recent refinement, Cationic Charge Ratio for Soil Structural Stability (CROSSc). The chapter concludes with an exploration of emerging tools, such as machine learning, dielectric sensing, and remote sensing, that enable rapid and spatially extensive diagnostics. Together, these approaches provide a robust framework for classifying dispersive soils and informing appropriate land and infrastructure management strategies.

7.2. TRADITIONAL PHYSICAL TEST METHODS FOR DISPERSIVE SOILS

7.2.1 Crumb test (Emerson test)

The crumb test, originally developed by Emerson (1967), is a widely used field and laboratory method for the rapid visual assessment of dispersive potential. An air-dried soil aggregate (typically 5–10 mm in diameter) is placed into distilled water or a dilute sodium hydroxide solution and observed for the release of clay colloids over 5–10 min. The reaction is categorised into four grades:

- Grade 1: No dispersion; the crumb may slake, but the water remains clear
- Grade 2: Slight cloudiness around the crumb; generally considered non-dispersive
- Grade 3: Moderate turbidity with visible clay halo; indicative of moderate dispersion
- Grade 4: High turbidity with complete clouding of water; strongly dispersive.

The crumb test is included in standard procedures such as ASTM D6572-21 (ASTM International 2021). Its appeal lies in its simplicity and low cost. However, its limitations include observer subjectivity, the influence of sample preparation (e.g. aggregate sieving), and variable responses due to clay mineralogy or organic matter. Kaolinitic soils, for example, may yield false negatives due to their relatively low swell potential. Consequently, the results of the crumb test should be interpreted in combination with chemical indices or results of quantitative tests.

7.2.2 Loveday and Pyle test

Loveday and Pyle (1973) refined the crumb test to include both dry aggregates and remoulded wet soil cubes, scoring dispersion after 2 and 20 h on a scale of 0–4. The four dispersion scores are summed to give a maximum possible value of 16. This semiquantitative approach allows differentiation between spontaneous (osmotic) and induced (mechanical) dispersion as follows:

- dry aggregates: reflect dispersion caused by osmotic forces
- remoulded cubes: indicate additional effects of mechanical stress.

A threshold dispersion index of 8 has been linked to ESP values around 6% and significant loss of hydraulic conductivity, thus offering predictive utility for soil amelioration decisions.

7.2.3 Aggregate stability in water test

The aggregate stability in water test (ASWAT), developed by Field *et al.* (1997), assesses dispersion using both air-dried and remoulded samples. Visual scoring is performed at 10 min and 2 h, with total scores for air-dried aggregates and remoulded samples calculated as follows:

- air-dry aggregates: total score = (10-min score + 2-h score) + 8
- remoulded samples: total score = (10-min score + 2-h score)

Total scores for air-dried aggregates range from 9 to 16, whereas those for remoulded samples range from 0 to 8.

A simplified protocol uses only the 10-min score. A total score >6 is considered indicative of dispersivity requiring amelioration.

7.2.4 Double hydrometer test

The double hydrometer test provides a quantitative measure of dispersive potential by comparing the proportion of clay that naturally disperses in water (without chemical dispersants) to the total clay content determined with dispersants (Decker and Dunnigan 1977). Two parallel hydrometer analyses are conducted:

1. Standard test with dispersant (e.g. sodium hexametaphosphate) and agitation: this test determines total clay (<0.005 mm).
2. Modified test without dispersant, using only deionised water and minimal agitation: this test measures natural dispersion.

The dispersion percentage (also called the dispersion ratio) is calculated as follows:

Dispersion percentage = (Clay from no-dispersant test/Total clay from full-dispersant test) × 100

Dispersion percentage values range from 0% (fully flocculated) to 100% (fully dispersive). In practice, values from >30% to 50% indicate dispersive or marginally dispersive potential and values <30% typically indicate non-dispersive potential.

Sherard *et al.* (1976) proposed a dispersion percentage >30% as a threshold for concern, and ASTM D4221 (ASTM International 2018) reports that 85% of known dispersive clays show >35% dispersion.

The double hydrometer test is more objective than visual tests but can underestimate dispersion in soils that require induction periods or contain cementing agents. Hence, it is best interpreted alongside complementary tests.

7.2.5 Pinhole test

The pinhole test (ASTM D4647; AS 1289.3.8.3; ASTM International 2020*a*) directly simulates internal erosion by passing water through a compacted soil sample via a 1-mm hole. Classification is based on flow clarity and pinhole expansion across multiple heads (50–1020 mm):

- ND1: non-dispersive (clear flow at all heads)
- D1: highly dispersive (cloudy flow and rapid enlargement at low head).

7.2.5.1 Interpreting and combining physical tests

Given the inherent limitations of individual tests, best practice involves using multiple physical methods to build a weight-of-evidence assessment of soil dispersivity. Each method offers distinct strengths and blind spots; combining them enhances diagnostic reliability through cross-validation.

For example:

- the crumb test may fail to detect dispersion in soils requiring mechanical agitation or delayed wetting responses
- the pinhole test may classify soils with high clay cohesion as non-dispersive, even if their chemistry suggests instability
- the double hydrometer test may under-report dispersion in soils that are cemented or slow to disperse without mechanical input
- ASWAT results may vary depending on sample remoulding technique and moisture conditions.

Cross-checking results (e.g. using the crumb test in tandem with the pinhole or double hydrometer test) can reveal consistent indicators of dispersivity. Concordant results (e.g. multiple tests confirming dispersion) offer robust evidence. Where test outcomes diverge, professional judgement is required, typically giving greater weight to methods that best simulate field conditions (e.g. the pinhole test), while incorporating chemical data to elucidate underlying causes.

Modern guidelines and research emphasise that physical test results, such as those from the pinhole, double hydrometer, crumb, or aggregate immersion tests, should not be interpreted in isolation. Reliable diagnosis of dispersive behaviour requires that:

- tests are conducted on samples replicating field conditions, particularly in terms of natural moisture content, bulk density, and compaction
- procedures follow standardised and repeatable protocols, including uniform aggregate sizes, consistent water quality, and defined observation durations

- results are evaluated within the broader context of site characteristics, including landscape position, field erosion features, known hydraulic behaviour, and land-use history.

This integrative approach improves diagnostic confidence and aligns with best practices recommended by soil science authorities and geotechnical standards (e.g. Northcote and Skene 1972; Sherard *et al.* 1976; Rengasamy *et al.* 1984; ASTM International 2018, 2020).

Operator variability also plays a significant role, especially in tests involving visual interpretation, such as the crumb test and ASWAT. Differences in aggregate selection, timing, and water quality can affect outcomes. Standardised protocols and observer training are essential to minimise inconsistencies.

In summary, the crumb, ASWAT, pinhole, and double hydrometer tests remain core tools for diagnosing soil dispersivity. When carefully implemented and interpreted as part of a multimethod approach, they provide a practical and reliable foundation for assessment. Their true value emerges when used in conjunction with chemical indicators that confirm sodicity and electrolyte conditions, topics addressed in the next section.

7.3 CHEMICAL INDICES AND SOIL PROPERTIES RELATED TO DISPERSION

The propensity of soils to disperse is fundamentally governed by chemical interactions within the solid and solution phases, particularly the presence and proportion of exchangeable sodium. A suite of chemical indices has been developed to assess dispersion risk, ranging from traditional metrics such as ESP and SAR to advanced indices including CROSS, its electrochemical refinement CROSSc, and the net dispersive charge (NDC). When interpreted alongside physical tests, these indices provide a robust framework for diagnosing soil dispersivity.

7.3.1 Exchangeable sodium percentage

ESP is defined as the proportion of a soil's cation exchange capacity (CEC) that is occupied by exchangeable sodium (Na^+):

$$ESP\ (\%) = (\text{Exchangeable } Na^+/CEC) \times 100$$

Traditionally, an ESP greater than 15% has been used to classify soils as sodic. However, research on Australian soils has demonstrated that dispersion can occur at much lower ESP values, particularly when electrolyte concentrations in the soil solution are low (McKenzie and Bryce 2023). For example, ~17% of soils with ESP less than 6% still exhibit dispersion, whereas ~3% of soils with higher ESP values remain structurally stable, often due to the presence of mitigating factors such as elevated salinity or calcium carbonate content. These findings underscore the limitations of ESP as a sole diagnostic criterion. Dispersion is influenced not only by sodium levels but also by the salinity of the soil solution, the type of clay minerals present (e.g. smectite versus kaolinite), soil pH, and the presence of stabilising agents like lime. To improve predictive capacity, the electrochemical stability index (ESI) has been introduced, defined as:

$$ESI = EC/(ESP/100)$$

where EC is electrical conductivity.

The ESI accounts for the balancing role of soluble salts, with low ESI values indicating conditions where sodium-induced dispersion is more likely, particularly following salt leaching from rainfall.

7.3.2 Sodium adsorption ratio

The SAR estimates the relative concentration of sodium to calcium and magnesium in soil solution or water extracts:

$$SAR = \mathrm{Na}/\sqrt{([\mathrm{Ca} + \mathrm{Mg}]/2)}$$

with all concentrations in $\mathrm{mmol_c\ L^{-1}}$.

Although SAR is primarily used to evaluate irrigation water quality, it also serves as a proxy for ESP through empirical correlations, such as:

$ESP \approx 1.95 \times SAR - 1.8$ (a relationship that has been empirically derived for many Australian soils)

An SAR greater than 13 in irrigation water is generally associated with sodicity risks, and a soil solution SAR above 5–6 commonly corresponds with ESP values exceeding 6%. Nevertheless, SAR lacks sensitivity to total ionic strength. As such, it is often interpreted alongside EC or total cation concentration to delineate the boundaries between flocculation and dispersion. SAR versus EC phase diagrams remain standard tools in assessing soil structural stability.

7.3.3 Cation Ratio of Soil Structural Stability

Developed by Rengasamy and Marchuk (2011), the CROSS index builds on SAR by incorporating potassium and weighting cations according to their relative flocculating or dispersive effects:

$$CROSS = (C_{\mathrm{Na}} + 0.56C_{\mathrm{K}})/[(C_{\mathrm{Ca}} + 0.60C_{\mathrm{Mg}})/2]^{0.5}$$

This formulation, based on cation concentrations in the soil solution ($\mathrm{mmol_c\ L^{-1}}$), typically signals increased risk of structural instability when values exceed 1.

Compared with SAR, CROSS has demonstrated better predictive accuracy in systems affected by high potassium inputs or variable water quality (Rengasamy and Marchuk 2011).

7.3.4 Cationic Charge Ratio for Soil Structural Stability

CROSSc is a recent refinement proposed by Rengasamy and Tavakkoli (2025) to further enhance diagnostic precision. It builds upon CROSS by explicitly incorporating the total electrostatic charge of all major cations (Na^+, potassium (K^+) versus calcium (Ca^{2+}), magnesium (Mg^{2+})) and removing assumptions such as fixed K^+ weighting or linearity in ion interactions. Although still under development and not yet standardised, early evidence suggests that CROSSc provides improved predictive accuracy in complex or variable soil systems (Rengasamy and Tavakkoli 2025). It reflects ongoing advances in electrochemical modelling of soil structural behaviour.

7.3.5 Net dispersive charge

The NDC provides a mechanistic evaluation of dispersion risk by calculating the net effect of dispersive and flocculating forces based on weighted cation concentrations. These are defined as follows:

$$\text{Dispersive charge} = (\text{Ca}) + 1.7(\text{Mg}) + 25(\text{K}) + 45(\text{Na})$$

$$\text{Flocculating charge} = 45(\text{Ca}) + 27(\text{Mg}) + 1.8(\text{K}) + (\text{Na})$$

$$NDC = \text{Dispersive charge} - \text{Flocculating charge}$$

with cation values expressed in $\text{cmol}^+ \text{ kg}^{-1}$. A positive NDC suggests a high likelihood of dispersion, whereas a zero or negative value indicates structural stability. NDC is particularly useful for diagnosing soils with low electrolyte levels, mixed mineralogy, or where traditional indices provide conflicting results. Moreover, it assists in formulating remediation strategies, such as gypsum addition to increase calcium-mediated flocculation (Rengasamy *et al.* 2016).

7.3.6 Other relevant soil properties

In addition to quantitative indices, several inherent soil characteristics significantly influence the potential for dispersion. Soils dominated by smectitic clays tend to be more dispersive than those rich in kaolinite due to their higher surface charge and swelling capacity. Smectites typically exhibit CECs ranging from 80 to 150 $\text{cmol}_c \text{ kg}^{-1}$, in contrast with 5–15 $\text{cmol}_c \text{ kg}^{-1}$ for kaolinites.

High soil pH, particularly above 9, can indirectly enhance dispersion by altering ion solubility and cation exchange equilibria. The presence of calcium carbonate (lime) in moderate amounts may mitigate sodicity by releasing Ca^{2+} into the soil solution, thereby promoting flocculation.

A comprehensive chemical characterisation of dispersive soils should include exchangeable cations (Na^+, K^+, Ca^{2+}, Mg^{2+}), EC, pH, CEC, and the lime or gypsum content. Multivariate indices that integrate these variables consistently outperform ESP alone in predicting dispersive behaviour.

7.3.7 Integrated interpretation and practical application

For practical soil management, chemical indices should not be interpreted in isolation. Diagnostic accuracy is significantly improved when they are used in conjunction with physical tests, such as the crumb, pinhole, or ASWAT tests. For instance, findings for a soil of ESP greater than 15% and dispersion in the pinhole test provide strong confirmation of sodicity. Conversely, if a soil with ESP below 6% shows dispersive behaviour, the findings may reflect low electrolyte concentration or a mineralogy rich in reactive clays. Thus, a holistic diagnostic framework involves triangulating data from chemical assessments, physical testing, and field observations. Increasingly, this approach is being augmented by emerging technologies, including artificial intelligence (AI)-based models and remote sensing, which are explored further in the next section.

7.4 EMERGING TECHNIQUES AND INTEGRATED APPROACHES IN DISPERSIVITY IDENTIFICATION

Advances in sensing technologies, data analytics, and computational tools have significantly transformed the identification of dispersive soils across diverse landscapes. These emerging approaches, which complement traditional physical and chemical tests, enable faster, spatially extensive, and often non-invasive assessments (Awais *et al.* 2023; Wadoux 2025). This section explores four key innovations: AI-based prediction models, dielectric and geophysical assessments, remote and proximal sensing, and integrated diagnostic frameworks. Collectively, these innovations enhance the accuracy and scalability of dispersive soil identification.

7.4.1 AI-based prediction models

The increasing availability of high-quality soil datasets, coupled with improvements in computational power, has facilitated the application of AI and machine learning in predicting soil dispersivity. These models are trained on databases containing soil samples with known dispersive behaviour and a wide range of measured properties, including grain size distribution, ESP, Atterberg limits, and clay mineralogy. Studies have shown that AI algorithms, including artificial neural networks and support vector machines, are capable of capturing the complex, non-linear relationships among variables that traditional single-parameter thresholds may miss (Olawade *et al.* 2024; Yaghoubi *et al.* 2024).

These tools have been used in a range of contexts, such as preliminary screening in geotechnical investigations, field prioritisation based on existing survey data, and the development of intuitive classification systems that assign risk levels (e.g. non-dispersive, marginal, highly dispersive). Although these models do not replace the need for site-specific laboratory testing, they serve as powerful decision-support tools. To ensure their robustness, emphasis must be placed on the transparency of model architecture and the quality of training data (Minasny *et al.* 2024).

7.4.2 Dielectric and geophysical assessments

Dielectric measurements offer a non-destructive approach to evaluating soil structure and potential for dispersion. Techniques such as time-domain reflectometry and dielectric spectroscopy provide insights into how soils interact with electromagnetic fields across a range of frequencies (Ngui and Lin 2019). Research indicates that dispersive sodium-dominated clays exhibit distinct dielectric signatures due to alterations in diffuse double-layer behaviour. In addition, microwave-range dielectric dispersion is linked to differences in pore structure and clay alignment.

Geophysical methods such as electromagnetic induction and electrical resistivity tomography further enhance dispersivity diagnostics by enabling in-field mapping of soil EC (Paz *et al.* 2024). These techniques help identify subsurface conditions associated with low salinity and high sodicity, factors conducive to dispersion. The non-invasive nature of these methods and their ability to provide spatially resolved data make them particularly well suited for large-scale assessments in agriculture, land development, and along infrastructure corridors. Although these techniques do not provide definitive diagnoses, they are effective for prioritising sampling zones and augmenting laboratory results (Moghadas *et al.* 2020; Farhat *et al.* 2022).

7.4.3 Remote and proximal sensing

Remote and proximal sensing technologies provide a means to detect landscape-level indicators of dispersive soils. Remote sensing, through satellite or unmanned aerial vehicle platforms, and proximal sensing, using yield monitors or in-field sensors, enable the detection of patterns associated with soil instability. Indicators such as persistent yield decline zones, vegetation stress captured via the normalised difference vegetation index, surface crusting or ponding, and salt efflorescence are increasingly used to infer dispersive behaviour (Kumar *et al.* 2024).

Integrated workflows have emerged where remote-sensed data are used to map candidate areas, followed by targeted ground-truthing and dispersivity testing. These data are then incorporated into geographic information systems alongside layers such as ESP maps, soil classifications, and erosion features to assess dispersivity risk at the landscape scale. Remote sensing is particularly valuable in precision agriculture and infrastructure planning, allowing the early identification and mitigation of dispersive soils (Fadl *et al.* 2024; Ambaru *et al.* 2025).

7.4.4 Integrated frameworks and standards

Modern soil diagnostic practices increasingly emphasise the integration of traditional and emerging tools within a structured framework (Kaplan *et al.* 2024). Updated standards and guidelines now recommend combining physical tests such as the crumb, pinhole, and double hydrometer tests with chemical indices like ESP and EC. Moreover, there is growing adoption of composite classification systems that categorise soils based on the integration of multiple lines of evidence.

For example, soils may be classified into dispersivity classes as follows:

- Class 1: non-dispersive (pinhole ND1–ND2, crumb Grade 1–2, ESP <6%)
- Class 2: moderately dispersive (pinhole ND3, crumb Grade 3, ESP 6–15%)
- Class 3: highly dispersive (pinhole D1–D2, crumb Grade 4, ESP >15%).

These classifications are supported by updates to key standards, including ASTM D6572 (crumb test; ASTM International 2020*b*), ASTM D4221 (double hydrometer; ASTM International 2018), ASTM D4647 (pinhole test; ASTM International 2020*a*), and enhancements to the AS 1289 series in Australia (Standards Australia 2020), which provide field-representative protocols for compacted soils. Such frameworks emphasise evidence triangulation to reduce diagnostic uncertainty and promote consistent, defensible soil classification.

7.4.5 From identification to management

Accurate identification of dispersive soils is the foundation for effective management strategies. The prevailing paradigm has shifted towards an 'identify-then-treat' model, where diagnostic outcomes inform timely intervention. For instance, a dispersive horizon detected through machine learning models may prompt early soil amendment or engineering design changes. Similarly, remote-sensed indicators may lead to stratified sampling, classification, and targeted amelioration measures at field or landscape scales.

The true value of these emerging techniques lies not only in their technological sophistication but also in their integration into practical, context-specific workflows. When used collaboratively, these tools provide robust, scalable, and adaptive systems for managing the challenges posed by dispersive soils.

From both engineering and agricultural perspectives, the identification of dispersive soils must be approached as a multidimensional challenge involving physical, chemical, and, increasingly, computational diagnostics. Dispersivity is not adequately captured by conventional geotechnical tests alone, such as particle size distribution, Atterberg limits, or compaction characteristics. Similarly, chemical criteria, including high ESP or SAR, are indicative of potential dispersion but do not provide conclusive evidence of field behaviour.

The integration of physical and chemical testing remains fundamental, not only for accurate diagnosis but also for informing remediation strategies that are tailored to soil type, condition, and landscape context. The deployment of machine learning models, such as artificial neural networks and support vector machines, presents new opportunities to capture complex, non-linear interactions among soil properties (Zhang *et al.* 2024). However, these models must be trained using high-quality, scenario-representative datasets to ensure reliability and transferability (Topçu and Seyrek 2024).

Emerging sensing technologies offer further diagnostic potential. Dielectric property measurements, including dielectric constant and loss, reflect changes in soil structure and water–clay interactions. These properties, increasingly assessed via microwave-based remote sensing and dielectric spectroscopy, can help map dispersivity across broad landscapes. In agricultural settings, it is essential to ensure that once dispersive soils are treated, they retain or recover their functional capacity. Post-treatment monitoring must include indicators such as hydraulic conductivity, active porosity, and dispersed clay content, in addition to chemical parameters like CROSSc and soil pH, which guide amendment effectiveness and inform crop health outcomes.

Ultimately, the identification and management of dispersive soils are no longer constrained to isolated, one-off tests. Instead, an evolving toolkit of laboratory, geospatial, and AI-enabled methods supports integrated, multiscale diagnostics. This approach is central to mitigating dispersion risks, enhancing infrastructure stability, and sustaining productive and resilient agroecosystems in dispersive soil landscapes.

7.5 REFERENCES

Ambaru B, Manvitha R, Madas R (2025) Synergistic integration of remote sensing and soil metagenomics data: advancing precision agriculture through interdisciplinary approaches. *Frontiers in Sustainable Food Systems* **8**, 1499973. doi:10.3389/fsufs.2024.1499973

ASTM International (2018) 'ASTM D4221-18. Standard test method for dispersive characteristics of clay soil by double hydrometer.' Available at https://doi.org/10.1520/D4221-18 [verified 30 July 2025].

ASTM International (2020*a*) 'ASTM D4647/D4647M-13(2020). Standard test methods for identification and classification of dispersive clay soils by the pinhole test.' Available at https://doi.org/10.1520/D4647_D4647M-13R20 [verified 30 July 2025].

ASTM International (2020*b*) 'ASTM D6572-20. Standard test methods for determining dispersive characteristics of clayey soils by the crumb test.' Available at https://doi.org/10.1520/D6572-20 [verified 30 July 2025].

ASTM International (2021) 'ASTM D6572-21. Standard test methods for determining dispersive characteristics of clayey soils by the crumb test.' Available at https://store.astm.org/d6572-21.html [verified 17 July 2025].

Awais M, Naqvi SMZA, Zhang H, Li L, Zhang W, Awwad FA, Ismail EAA, Khan MI, Raghavan V, Hu J (2023) AI and machine learning for soil analysis: an assessment of sustainable agricultural practices. *Bioresources and Bioprocessing* **10**, 90. doi:10.1186/s40643-023-00710-y

Decker RS, Dunnigan LP (1977) Development and use of the soil conservation service dispersion test. In 'Proceedings of symposium on dispersive clays, related piping and erosion in geotechnical projects'. ASTM Special Publication 623. (Eds JL Sherard, RS Decker) pp. 94–109. (ASTM International)

Emerson W (1967) A classification of soil aggregates based on their coherence in water. *Soil Research* **5**, 47–57. doi:10.1071/SR9670047

Fadl ME, AbdelRahman MAE, El-Desoky AI, Sayed YA (2024) Assessing soil productivity potential in arid region using remote sensing vegetation indices. *Journal of Arid Environments* **222**, 105166. doi:10.1016/j.jaridenv.2024.105166

Farhat I, Farrugia L, Bonello J, Sammut C, Persico R (2022) Measuring the dielectric properties of soil: a review and some innovative proposals. In 'Instrumentation and measurement technologies for water cycle management'. (Eds A Di Mauro, A Scozzari, F Soldovieri) pp. 485–509. (Springer International Publishing: Cham)

Field DJ, McKenzie DC, Koppi AJ (1997) Development of an improved Vertisol stability test for SOILpak. *Soil Research* **35**, 843–852. doi:10.1071/S96118

Kaplan S, Ropelewska E, Günaydın S, Sabancı K, Çetin N (2024) Machine learning and computer vision technology to analyze and discriminate soil samples. *Scientific Reports* **14**, 19945. doi:10.1038/s41598-024-69464-7

Kumar D, Pradhan AK, Jain R, Kumar V, Murmu S, Samal I, Chaurasia HS (2024) Remote sensing in precision agriculture: current status and applications. In 'Artificial intelligence and smart agriculture: technology and applications.' (Eds K Pandey, NL Kushwaha, CB Pande, KG Singh) pp. 23–41. (Springer Nature Singapore: Singapore)

Loveday J, Pyle JC (1973) 'The Emerson dispersion test and its relationship to hydraulic conductivity.' (CSIRO Division of Soils: Melbourne, Vic.)

McKenzie D, Bryce A (2023) Dispersive soil manual. Managing dispersive soils: practicalities and economics – Northern region. Grains Research and Development Corporation. Available at https://grdc.com.au/resources-and-publications/all-publications/publications/2023/dispersive-soil-manual [verified 14 July 2025].

Minasny B, Bandai T, Ghezzehei TA, Huang Y-C, Ma Y, McBratney AB, Ng W, Norouzi S, Padarian J, Rudiyanto, Sharififar A, Styc Q, Widyastuti M (2024) Soil science-informed machine learning. *Geoderma* **452**, 117094. doi:10.1016/j.geoderma.2024.117094

Moghadas D, Behroozmand AA, Christiansen AV (2020) Soil electrical conductivity imaging using a neural network-based forward solver: applied to large-scale Bayesian electromagnetic inversion. *Journal of Applied Geophysics* **176**, 104012. doi:10.1016/j.jappgeo.2020.104012

Ngui YJ, Lin C-P (2019) 'Broadband complex dielectric characterization of soils by time domain reflectometry.' (Springer Singapore: Singapore)

Olawade DB, Wada OZ, Ige AO, Egbewole BI, Olojo A, Oladapo BI (2024) Artificial intelligence in environmental monitoring: advancements, challenges, and future directions. *Hygiene and Environmental Health Advances* **12**, 100114. doi:10.1016/j.heha.2024.100114

Paz MC, Castanheira NL, Paz AM, Gonçalves MC, Monteiro Santos F, Farzamian M (2024) Comparison of electromagnetic induction and electrical resistivity tomography in assessing soil salinity: insights from four plots with distinct soil salinity levels. *Land (Basel)* **13**, 295. doi:10.3390/land13030295

Rengasamy P, Marchuk A (2011) Cation ratio of soil structural stability (CROSS). *Australian Journal of Soil Research* **49**, 280–285.

Rengasamy P, Tavakkoli E (2025) Cationic charge ratio (CROSSc) as an index for dispersive soils and irrigation water quality: advancing beyond SAR and CROSS. *Earth Critical Zone* **2**, 100025. doi:10.1016/j.ecz.2025.100025

Rengasamy P, Tavakkoli E, McDonald GK (2016) Exchangeable cations and clay dispersion: net dispersive charge, a new concept for dispersive soil. *European Journal of Soil Science* **67**, 659–665. doi:10.1111/ejss.12369

Sherard JL, Dunnigan LP, Decker RS (1976) Identification and nature of dispersive soils. *Journal of the Geotechnical Engineering Division* **102**(4). doi:10.1061/AJGEB6.000025

Standards Australia (2020) 'AS 1289.6.8.4–2020: Methods of testing soils for engineering purposes – soil chemical tests – determination of emerson class number of a soil – modified Emerson class number test.' (Standards Australia: Sydney, NSW)

Topçu S, Seyrek E (2024) A new approach for identification of dispersivity class of soils by combining physical and chemical tests. *Physics and Chemistry of the Earth Parts A/B/C* **135**, 103685. doi:10.1016/j.pce.2024.103685

Vakili AH, Salimi M, Keskin İ, Jamalimoghadam M (2024) A systematic review of strategies for identifying and stabilizing dispersive clay soils for sustainable infrastructure. *Soil & Tillage Research* **239**, 106036. doi:10.1016/j.still.2024.106036

Wadoux AMJ-C (2025) Artificial intelligence in soil science. *European Journal of Soil Science* **76**, e70080. doi:10.1111/ejss.70080

Yaghoubi E, Yaghoubi E, Khamees A, Vakili AH (2024) A systematic review and meta-analysis of artificial neural network, machine learning, deep learning, and ensemble learning approaches in field of geotechnical engineering. *Neural Computing & Applications* **36**, 12655–12699. doi:10.1007/s00521-024-09893-7

Zhang X, Liu Z, Han Y (2024) Progress towards the identification and improvement of dispersive soils: a review. *European Journal of Soil Science* **75**, e70002. doi:10.1111/ejss.70002

8

Knowledge gaps and future perspectives in dispersive soils research

8.1 INTRODUCTION

Dispersive soils, characterised by the breakdown and migration of clay particles when wetted, represent a widespread and insidious threat to land productivity, water quality, and infrastructure stability. Often linked to high sodicity and low electrolyte concentrations, these soils exhibit structural collapse, erosion, surface sealing, and tunnel formation across a range of climatic and land use contexts. Globally, dispersive and sodic soils are estimated to affect more than 5.8 million square kilometres, making them one of the most extensive forms of soil degradation. Yet despite their scope and severity, dispersive soils remain under-recognised in policy frameworks and under-researched in relation to other land degradation drivers like salinisation or acidification.

This chapter outlines the critical knowledge gaps and innovation priorities that must be addressed to manage dispersive soils effectively. Building on the classification, mechanisms, and management strategies covered in earlier chapters, we move from a national to a global perspective, drawing on recent advances in diagnostic tools, biological restoration, eco-engineering, and digital monitoring. We also examine the interlocking roles of climate change, salinity shifts, land use intensification, and poor water quality in exacerbating dispersion risks.

Importantly, dispersive soils demand a multidisciplinary response. They influence and are influenced by physical, chemical, biological, and hydrological processes, and their management must account for socioeconomic realities on farms, in urban planning, and across natural ecosystems. As such, this chapter embraces a systems perspective, weaving in case studies from agriculture, infrastructure, and conservation to highlight the real-world implications of dispersion.

We structure the chapter around 10 core themes, reflecting both diagnostic and management dimensions:

- challenges in identifying dispersive soils under field conditions
- disruptions to soil–plant–water interactions across climates and land uses
- acceleration of dispersion via salinisation, climate extremes, and human activity
- gaps and limitations in current amelioration practices
- emerging biological and microbial solutions
- technological innovations in sensing, mapping, and predictive modelling
- interdisciplinary integration across soil science, engineering, ecology, and economics
- regional research needs tailored to different soil–climate systems
- policy innovations, governance gaps, and incentive mechanisms
- global collaboration and future directions in dispersive soil research.

By critically examining these themes, we aim to position dispersive soils not as a fixed agronomic constraint, but rather as a dynamic frontier where scientific innovation, policy design, and practical stewardship can converge to reverse degradation trends. The goal is to transform dispersive soils from neglected liabilities into productive, resilient, and ecologically functional landscapes.

8.2 CHALLENGES IN THE IDENTIFICATION AND CLASSIFICATION OF DISPERSIVE SOILS

Accurate identification and classification of dispersive soils remain foundational yet persistently difficult tasks. Despite decades of research, current diagnostic methods fail to consistently predict real-world soil behaviour, leading to misclassification, failed amelioration, and infrastructure vulnerability. Traditionally, dispersivity has been diagnosed using a suite of physical and chemical indicators, such as the Emerson crumb test, exchangeable sodium percentage (ESP), and the sodium adsorption ratio (SAR). However, these tools often fall short, especially under field conditions, where dispersion is influenced by a complex interplay of mineralogy, water chemistry, and environmental context.

Field methods such as the Emerson crumb test offer a rapid visual assessment of aggregate stability, but their reliability is compromised in soils with high kaolinite content or organic coatings that suppress visible dispersion. Likewise, ESP thresholds, typically >15% for sodicity classification, do not consistently predict dispersion. Numerous studies have documented dispersive behaviour in soils with ESP values as low as 3–5%, particularly when electrolyte concentrations are depleted through rainfall or irrigation (Shainberg *et al.* 1981; Rengasamy 2018). Conversely, some soils with high ESP values (>25%) remain aggregated due to protective mineralogical or saline conditions. These inconsistencies point to a deeper issue: sodicity is a necessary but not sufficient condition for dispersion.

The SAR index, although useful in water quality assessment, fails to account for ionic strength and actual dispersion potential *in situ*. More recently, the Cation Ratio of Soil Structural Stability (CROSS) index (Rengasamy and Marchuk 2011) and net dispersive charge concept (Rengasamy *et al.* 2016) have emerged as more robust indicators. Unlike ESP or SAR, these concepts incorporate the flocculating and dispersing effects of all major cations (Ca^{2+}, Mg^{2+}, Na^{+}, K^{+}) along with electrolyte concentration, offering a more nuanced prediction of clay stability. Application of CROSS in semi-arid Vertosols in India and irrigated Sodosols in China has shown superior correlation with field dispersion and erosion compared with legacy indices (Vakili *et al.* 2024).

Despite these advances, most national and international soil classification systems, including the US Department of Agriculture Soil Taxonomy and the World Reference Base for Soil Resources, still rely on static thresholds like ESP and fail to incorporate dynamic indicators such as CROSS, zeta potential, or dispersion kinetics under wetting conditions. Moreover, current field practices rarely integrate chemical diagnostics with mineralogical data (e.g. smectite versus kaolinite), which significantly affects dispersivity. This gap is particularly problematic in large-scale developments, such as road construction or dam siting, where undetected dispersive subsoils can lead to catastrophic failure.

Emerging approaches are beginning to address these challenges. Machine learning and artificial intelligence (AI) models, trained on large soil datasets, now show promise in predicting dispersion based on a combination of physical, chemical, and spatial variables. Neural networks, decision trees, and support vector machines have been used to develop dispersion risk maps in Australia and the US, outperforming traditional threshold-based models (Wu *et al.* 2018). These tools can identify complex non-linear interactions and are particularly valuable when paired with proximal sensing technologies and digital soil mapping platforms.

In addition, portable field instruments are improving in speed and resolution. For example, real-time turbidity tests, dispersion ratio sensors, and zeta potential probes are being developed to provide on-site dispersion risk assessments. When coupled with mobile apps or cloud-based data platforms, these tools may soon enable decision-makers (farmers, engineers, planners) to rapidly diagnose dispersive risk and select appropriate interventions on the spot.

8.2.1 Knowledge gaps

Despite growing toolsets, several persistent knowledge gaps limit progress; specifically:

- There is no universally accepted integrated framework that combines chemical, mineralogical, physical, and environmental factors into a single predictive system for dispersive behaviour.
- The CROSS index, although promising, is not yet widely adopted and lacks calibration across diverse climates and soil types.
- AI-based models depend on high-quality, spatially resolved datasets, which are unavailable in many developing countries or remote regions.
- Field diagnostics remain too empirical or subjective, often based on visual judgement without quantitative validation.
- Most diagnostics fail to account for temporal dynamics, such as seasonal salinity fluctuations or biological activity, that modulate dispersive behaviour over time.

8.2.2 Future research directions

To advance dispersive soil diagnostics, a multitiered innovation agenda is required to:

- expand and calibrate advanced indices like CROSS across a range of geographies and soil types; this includes developing decision thresholds tailored to local conditions (e.g. rainfall regimes, clay mineralogy)
- integrate machine learning into national soil survey and digital soil mapping frameworks to generate spatially explicit dispersion risk models
- develop hybrid diagnostic protocols that combine laboratory tests (e.g. zeta potential, electrolyte titrations) with remote sensing and field sensors for rapid triangulation of soil behaviour
- create practical field kits with visual turbidity and electrical conductivity (EC)/sodium (Na) ratio tools, connected via mobile platforms to cloud-based advisory systems
- promote interdisciplinary collaboration between soil scientists, geotechnical engineers, computer scientists, and land managers to co-design field-applicable diagnostic systems
- include local knowledge and farmer observations in dispersion diagnostics, particularly in poorly mapped regions (participatory mapping of gully-prone areas can help ground-truth digital risk models).

Ultimately, effective identification of dispersive soils must move beyond one-size-fits-all metrics and embrace dynamic, location-specific, and multidisciplinary diagnostics. Doing so will unlock more timely, cost-effective, and scalable management strategies, critical for reducing erosion, preserving productivity, and protecting infrastructure in vulnerable landscapes.

8.3 SOIL–PLANT–WATER–CLIMATE INTERACTIONS IN DISPERSIVE SOILS

Dispersive soils significantly alter the interactions between soil structure, water dynamics, and plant development. When dispersed clays are wetted, structural collapse results in the formation of dense surface crusts and impermeable subsoil layers. These changes impair infiltration, restrict gas exchange, and limit root penetration, creating multiple agronomic constraints. Surface sealing impedes seedling emergence, whereas compacted subsoils reduce root access to water and nutrients. These effects are further compounded by water loss through run-off, resulting in moisture stress even in areas with adequate rainfall or irrigation.

Although dispersion is fundamentally a physicochemical phenomenon, its severity and agricultural impact are strongly modulated by climate and land use. In arid and semi-arid regions, salts accumulate at the surface during dry periods and can temporarily stabilise aggregates by promoting flocculation. However, upon the onset of rainfall or irrigation, these salts may be rapidly leached, triggering sudden dispersion. This behaviour has been observed in Vertosols and duplex soils in central Australia, West Asia, and Sub-Saharan Africa. In humid and sub-humid climates, continuous leaching of electrolytes heightens dispersion risk. Dispersive topsoils in these zones, such as in South-east Asia, Latin America, and parts of eastern Australia, readily form seals and experience gully erosion during heavy rainfall events. Monsoonal and tropical wet–dry climates present an additional pattern: salt accumulation in the dry season is followed by dispersive collapse in the wet season. This cyclical behaviour contributes to severe erosion and patchy vegetation cover in northern Australia and parts of South and South-east Asia.

The effect of land use and management on dispersive soils is equally profound. Undisturbed soils under natural vegetation often maintain structure due to organic inputs and biological activity. However, conversion to cropping, pasture, or infrastructure development frequently exposes these soils to physical disturbance and the removal of stabilising agents. Tillage and compaction can directly break down macroaggregates, whereas poor irrigation practices, particularly with sodium-rich water, exacerbate sodicity and lead to subsoil dispersion. Irrigated cropping systems in California's Central Valley and the Indo-Gangetic plains of South Asia have experienced a decline in soil hydraulic conductivity and crop performance due to such processes.

Not all land uses exacerbate dispersion. Systems that maintain root cover, reduce disturbance, and build organic matter, such as no-till, cover cropping, pasture rotation, or agroforestry, can support biological aggregation. Certain plant species with deep, fibrous roots or exudate profiles conducive to aggregation (e.g. lucerne, vetiver, saltbush) may enhance porosity and improve the resilience of dispersive soils. However, many commercial crops lack the traits needed to thrive in such environments, resulting in shallow rooting, waterlogging, and yield loss. In highly dispersive soils of eastern Australia and India, studies have reported cereal yield reductions of 30–70% under conventional management (McKenzie and Bryce 2023; Uddin *et al.* 2025).

8.3.1 Knowledge gaps

Several critical knowledge gaps limit our understanding of soil–plant–water interactions in dispersive environments. First, the thresholds at which dispersive behaviour manifests under different climatic conditions remain poorly defined. Although general associations between ESP, EC, and rainfall intensity exist, precise and regionally relevant dispersion thresholds are lacking. The effects of seasonal variation in moisture, salinity, and microbial activity on dispersion risk have not been adequately quantified. Moreover, hydrological models such as Agricultural Production System sIMulator (APSIM; Holzworth *et al.* 2014), Soil and Water Assessment Tool (SWAT; Arnold *et al.* 1998), and Water Erosion Prediction Project (WEPP; Laflen *et al.* 1991) do not currently account for dynamic dispersion-related changes in soil structure, limiting their ability to simulate infiltration decline, run-off generation, and yield loss in dispersive soils.

Plant-level responses are also poorly characterised. There is limited understanding of the root traits (e.g. architecture, exudation, or microbial associations) that enable some species to function better in dispersive conditions. Rhizosphere-mediated processes that influence clay flocculation or aggregate stabilisation remain largely unexplored. Although anecdotal and observational data suggest that cover crops, perennials, and deep-rooted species can mitigate dispersion impacts, few systematic studies have examined these relationships over time, across soil types, or in different rainfall zones.

Management practices offer further uncertainties. Although reduced tillage, stubble retention, and cover cropping are believed to support soil structure, their specific effects on dispersive behaviour, erosion risk, and crop productivity have not been studied in a mechanistic or comparative framework. There is also a lack of cross-climate studies assessing whether management strategies effective in one region (e.g. southern Australia) translate to other contexts (e.g. West Africa or

Central Asia). Finally, there are minimal data on the economic viability or adoptability of such strategies in different farming systems, particularly in smallholder or resource-constrained settings.

8.3.2 Future research directions

Addressing these knowledge gaps will require interdisciplinary, climate-responsive, and systems-based approaches. Long-term, multilocation field experiments should be established across arid, humid, and transitional zones to monitor how dispersive soils respond to different rainfall patterns, land uses, and cropping systems. These trials should quantify infiltration rates, erosion losses, biological activity, root development, and yield responses under both untreated and managed conditions.

Process-based hydrological models must be adapted to simulate dispersion-related changes in soil hydraulic conductivity and structural stability. Incorporating dispersion modules into models such as APSIM or SWAT would improve their relevance for risk assessment and land management decision-making. Sensor-based tools, including soil moisture tension probes, salinity sensors, and unmanned aerial vehicle-mounted normalised difference vegetation index and thermal cameras, should be deployed to monitor real-time changes in soil behaviour and crop response in dispersive landscapes.

Future research should also focus on identifying root traits and plant–microbe interactions that confer tolerance to dispersive conditions. This includes exploring the potential of rhizosphere engineering, biopriming, and crop rotations that integrate deep-rooted, aggregation-promoting species. Breeding programs may benefit from including root functional traits, such as exudate profiles, root length density, and hydraulic conductivity, as selection criteria in dispersive environments.

In terms of management innovation, region-specific packages that combine physical, chemical, and biological interventions must be developed and tested. For example, in semi-arid zones, strategies may focus on water conservation and surface stability; in humid zones, drainage and erosion control will be critical. These strategies must be evaluated not only for agronomic performance but also for their cost-effectiveness, scalability, and social acceptability in diverse farming systems.

Ultimately, managing soil–plant–water interactions in dispersive soils will depend on integrating scientific insight with practical innovation. By closing key knowledge gaps and developing climate-adapted, biologically informed strategies, we can mitigate the negative impacts of dispersion on crop production, water use efficiency, and long-term soil resilience.

8.4 ROLES OF SOIL SALINISATION, CLIMATE CHANGE, AND ANTHROPOGENIC PRESSURES IN ACCELERATING DISPERSION

Dispersive soils are not static in their behaviour or spatial extent. Rather, they evolve in response to a confluence of environmental, climatic, and anthropogenic drivers. Among these, soil salinisation, climate change, and human-induced land degradation are key processes that both reveal and intensify the expression of dispersive behaviour across landscapes. In many regions, soils that were once marginally sodic or stable under natural vegetation have become highly dispersive due to shifts in salinity regimes, irrigation practices, vegetation removal, and hydrological modification.

At the core of this transformation lies the dynamic relationship between sodicity and salinity. Although sodicity, characterised by a high proportion of exchangeable sodium ions, is a primary cause of dispersion, the presence of soluble salts (i.e. salinity) can temporarily suppress dispersive behaviour by maintaining electrolyte concentrations above critical flocculation thresholds. However, when these salts are removed, through rainfall, leaching, or drainage, the underlying sodic condition is unmasked, and dispersion is triggered. This phenomenon is widely documented in irrigation districts where salt-affected soils initially exhibit structural stability, only to collapse structurally following leaching or reclamation efforts. In many cases, efforts to improve plant growth through salt leaching inadvertently exacerbate dispersion, causing erosion, crusting, and reduced infiltration.

Climate change is anticipated to accelerate these processes. Shifts in rainfall intensity, frequency, and seasonal distribution are already altering soil–water–salt dynamics in vulnerable regions. High-intensity rainfall delivers large volumes of low-electrolyte water, increasing the potential for clay dispersion and erosion, particularly in soils with low organic matter and poor structural resilience. Conversely, prolonged drought concentrates salts at the surface, temporarily maintaining aggregation but creating a latent risk of dispersion upon rewetting. Rising temperatures and increased evaporative demand also enhance salt accumulation and sodification risk. In coastal zones, sea-level rise and saline water intrusion introduce additional complexity by increasing surface and subsoil salinity, which may initially mask sodicity but contribute to long-term instability when followed by rainfall or freshwater irrigation.

Anthropogenic pressures, including unsustainable irrigation, poor land clearing practices, overgrazing, and infrastructure development, compound these risks. In many regions, secondary sodification has arisen due to the long-term use of sodium-rich or bicarbonate-dominated irrigation water. Such practices alter the soil exchange complex, replacing calcium and magnesium with sodium and thereby increasing the soil's dispersive potential. The widespread application of sodium-based fertilisers, industrial effluents, and untreated wastewater further exacerbates this condition. In engineering and construction contexts, dispersive subsoils are often disturbed, relocated, or compacted during road and dam building, leading to tunnel erosion, piping, and embankment failure, especially when proper identification and management are lacking. Notable examples include road washouts in eastern Ethiopia and embankment failures in Queensland, Australia, both linked to the mismanagement of dispersive clays.

Vegetation loss, through overgrazing, deforestation, or fire, exposes soil surfaces to erosive forces, interrupts organic matter inputs, and removes biological mechanisms that promote aggregation. In overgrazed catchments such as the Burdekin Basin in northern Australia, these factors have contributed to extensive gully erosion, sediment transport, and ecosystem degradation, including impacts on the Great Barrier Reef (McKenzie and Bryce 2023). Similar patterns are emerging in parts of South Asia, the Middle East, and Sub-Saharan Africa, where dispersive soils are contributing to landscape fragmentation, reduced productivity, and sedimentation of water bodies (FAO 2015).

8.4.1 Knowledge gaps

Despite growing awareness of these converging threats, key knowledge gaps persist regarding the mechanisms, thresholds, and feedbacks that govern how salinisation, climate variability, and human activity interact to induce or intensify dispersion. First, the critical balance between salinity and sodicity remains poorly quantified. Although theoretical models suggest a minimum electrolyte concentration is needed to prevent clay dispersion at a given ESP, this threshold is not consistently defined across different soil textures, mineralogies, or climates. Empirical data from field settings are limited, particularly in regions experiencing rapid irrigation expansion or climate-induced salinity shifts.

Second, there is a lack of predictive modelling that incorporates climate change scenarios into the risk assessment of dispersive soil expansion. Most current soil degradation models do not include dynamic dispersion behaviour, nor do they simulate the structural consequences of salt leaching, episodic rainfall, or temperature-driven evaporative trends. As a result, projections of erosion, run-off, and productivity loss in dispersive landscapes remain uncertain.

The socioeconomic dimensions of dispersion risk are also underexplored. Although many studies document biophysical processes, few evaluate the human decision-making and policy frameworks that allow dispersive soil degradation to persist or escalate. For instance, the role of irrigation policy, water pricing, land tenure, and knowledge access in shaping farmer practices around sodicity management is rarely examined in the context of dispersion risk.

Finally, monitoring protocols for the early detection of salinity–sodicity transitions are underdeveloped. Few systems integrate remote sensing, soil sampling, and hydrological data to provide early warnings of when a stable sodic soil may become actively dispersive. Likewise, infrastructure planning rarely includes rigorous assessment of dispersion risk, despite the known consequences for embankment, road, and dam stability.

8.4.2 Future research directions

Addressing these challenges requires both technical innovation and institutional reform. From a technical perspective, more refined field-based thresholds for flocculation–dispersion transitions must be developed, integrating measurements of ESP, EC, SAR, CROSS values, and cation balance under real-world conditions. Coupling these thresholds with dynamic water balance modelling will allow better forecasting of when and where dispersion is likely to occur following seasonal rainfall or irrigation events.

Efforts should also focus on designing integrated reclamation strategies that manage both salinity and sodicity simultaneously. For example, leaching schedules must be paired with calcium amendments (e.g. gypsum, calcium chloride) to prevent structural degradation during desalinisation. In dryland and resource-limited regions, low-cost alternatives to conventional gypsum, such as by-products or *in situ* mineral weathering, should be explored for managing dispersion risk.

Climate-resilient land and water management will be essential. Predictive models that combine soil data with climate forecasts can help identify areas at heightened risk of dispersion under future rainfall regimes. These models can guide the placement of erosion control structures, gully remediation zones, and targeted revegetation efforts. Moreover, nature-based solutions, such as the use of salt-tolerant or deep-rooted species to improve aggregation and reduce run-off, should be developed and tested in different agroecological zones.

From a governance perspective, policies must better align soil degradation risk with land and water use regulations. Environmental impact assessments for new agricultural, urban, or infrastructure projects should include mandatory screening for dispersive soils. In irrigation schemes, water quality standards and soil monitoring should be enforced to prevent secondary sodification. Where dispersion is already active, regulatory and incentive frameworks should support amelioration, conservation agriculture, and restoration initiatives, particularly where public infrastructure or downstream ecosystems are affected.

Finally, a cross-disciplinary approach is required to integrate engineering, agronomy, hydrology, and policy perspectives. Collaborative monitoring systems, combining satellite data, *in situ* sensors, and participatory mapping, can support real-time detection and adaptive response. These platforms should be co-developed with farmers, land managers, and local institutions to ensure usability, scalability, and impact.

In sum, the accelerating influence of salinisation, climate change, and anthropogenic pressure on dispersive soil behaviour underscores the urgency of moving beyond reactive soil management. With integrated scientific, technological, and institutional action, it is possible to prevent the transformation of marginally sodic soils into fully degraded, structurally unstable landscapes and instead chart a path towards long-term resilience.

8.5 BIOLOGICAL AND MICROBIAL METHODS FOR SOIL RESTORATION: GAPS AND OPPORTUNITIES

Conventional approaches to managing dispersive soils have predominantly relied on chemical and physical interventions, most notably the application of calcium-based amendments (e.g. gypsum) and mechanical techniques such as deep ripping or surface reshaping. Although these approaches can offer short-term structural improvements, they often require repeated application, involve significant costs, and are difficult to scale across degraded or resource-limited landscapes. In this context, biological and microbial strategies are gaining attention as potentially cost-effective, sustainable, and ecologically integrative alternatives.

Biological interventions aim to enhance soil aggregation and structural stability through natural processes, including organic matter addition, rhizosphere activity, and microbial by-product formation. Recent studies have shown that the application of organic amendments, such as compost, manure, or crop residues, can stimulate microbial growth and lead to improvements in macroaggregate stability even in sodic or dispersive subsoils (Sale *et al.* 2021). The biological mechanisms at play include microbial exudation of extracellular polysaccharides, fungal hyphal entanglement, and enhanced root turnover, all of which contribute to the development of stable aggregates. In some cases, deep placement of organic matter in subsoil layers has been associated with increased porosity, deeper root penetration, and improved moisture retention, thereby enhancing crop performance (Fang *et al.* 2021; Tavakkoli *et al.* 2022).

Beyond organic matter application, the use of beneficial microbial consortia is emerging as a frontier in the biological remediation of dispersive soils. For instance, studies have demonstrated the potential of plant growth-promoting rhizobacteria, nitrogen-fixing bacteria, and arbuscular mycorrhizal fungi to stabilise soil structure through a combination of biochemical and biophysical pathways (Rilling *et al.* 2001; Mueller *et al.* 2024). Microbial-induced carbonate precipitation (Wang *et al.* 2023), where bacteria such as *Sporosarcina pasteurii* facilitate calcium carbonate formation, has shown promise in improving subsoil cohesion and reducing dispersivity in controlled environments (Carter *et al.* 2023; Fu *et al.* 2023). Biopolymers produced by bacteria and fungi, such as exopolysaccharides, also play an important role in enhancing particle aggregation and water retention (Netrusov *et al.* 2023). These natural binding agents may offer a biological analogue to synthetic soil conditioners, but with greater environmental compatibility.

Field applications of these approaches remain limited, but several pilot studies suggest that the co-application of microbial inoculants and organic substrates can improve hydraulic conductivity and reduce erosion risk in dispersive landscapes. For example, inoculating compost-treated dispersive soils with consortia of exudate-producing bacteria and arbuscular mycorrhizal fungi has led to improved infiltration rates and increased aggregate stability (Sahin *et al.* 2011; Trivedi *et al.* 2017). These results, although preliminary, indicate that biological approaches may offer a pathway towards functional restoration of dispersive soils without reliance on high-input or mechanistic strategies (Rog *et al.* 2025).

8.5.1 Knowledge gaps

Despite their potential, biological and microbial approaches to dispersive soil remediation remain under-researched and poorly understood at both mechanistic and applied levels. One critical gap lies in the limited identification and characterisation of the microbial taxa and functional traits responsible for stabilising dispersive clays. Most studies to date have focused on microbial abundance and diversity, rather than on the specific metabolic functions or biophysical interactions relevant to aggregate formation.

There is also a lack of clarity regarding the temporal dynamics and persistence of biologically induced improvements. Although structural gains have been observed in the short term following organic amendment or microbial inoculation, it is unclear whether these effects endure over multiple seasons or degrade as microbial activity declines. The extent to which microbial products such as biofilms, mucilage, or carbonate precipitates resist disturbance, drought, or salinity stress is yet to be determined.

Furthermore, the interactions between plant root systems and microbial communities in dispersive contexts remain largely unexplored. Certain plant species exude compounds that enhance microbial colonisation and aggregation, whereas others may degrade structure through root compaction or exudate composition. Identifying compatible plant–microbe combinations that enhance soil structure in sodic and dispersive environments is a research area with significant promise but limited empirical basis.

Another key gap is the paucity of field-scale demonstrations and the lack of techno-economic analysis. Although small-scale trials have shown potential, larger-scale and longer-term studies are needed to assess feasibility, cost-effectiveness, and replicability. Moreover, the influence of soil mineralogy, climatic conditions, and management history on biological amendment outcomes is not well understood. It remains unclear whether microbial strategies that succeed in temperate Vertosols, for example, would perform similarly in tropical or arid duplex soils.

8.5.2 Future research directions

To realise the potential of biological and microbial approaches for dispersive soil remediation, several research priorities must be addressed. First, there is a need for functional characterisation of soil microbial communities under dispersive conditions. This includes applying metagenomic, metatranscriptomic, and metabolomic tools to identify microbial pathways involved in aggregation, calcium mobilisation, and soil particle binding. Targeted isolation of exopolysaccharide-producing strains or carbonate-precipitating bacteria from native sodic environments may inform the development of tailored bioinoculant formulations.

Second, future work should focus on designing and testing microbial consortia in combination with organic substrates such as compost, biochar, or cover crop residues. These combinations may create synergistic effects that improve microbial establishment, nutrient cycling, and aggregation. Studies should explore both single-species and multispecies inoculants to determine optimal compositions under different soil and climatic conditions.

The role of plants as biological drivers of restoration must also be more fully integrated. Screening crop and pasture species for root traits that promote microbial colonisation, exudation of flocculating agents, and deep rooting into compacted dispersive layers can help inform the selection of species suitable for phytoremediation. Deep-rooted perennials and legumes, particularly those with known associations with arbuscular mycorrhizal fungi or plant growth-promoting rhizobacteria, should be evaluated for their capacity to contribute to long-term structural improvement (Sagar *et al.* 2021).

In parallel, research should investigate the scalability and practical implementation of these strategies. This includes evaluating the logistics of inoculant production and delivery, compatibility with existing farming systems, and cost–benefit analysis over time. Participatory trials with landholders and local extension networks can help refine application methods, assess farmer perceptions, and guide decision-support tools.

Finally, integrating biological approaches with digital and remote sensing technologies could facilitate the monitoring of treatment effectiveness. For example, changes in surface roughness, vegetation establishment, or infiltration rates could be detected via drone imagery or *in situ* sensors, providing timely feedback on biological remediation outcomes. Biological and microbial strategies for dispersive soil restoration remain in the early stages of development, yet they offer a compelling alternative or complement to traditional practices. With concerted research and cross-sector collaboration, these approaches could contribute significantly to restoring soil health, improving water use efficiency, and enhancing landscape resilience in sodic and dispersive environments.

8.6 EMERGING IMPACTS ON ECOSYSTEMS AND BIODIVERSITY

Dispersive soils not only present agronomic and engineering challenges but also pose substantial risks to terrestrial and aquatic ecosystems. Their effects extend beyond productivity constraints or erosion losses; they contribute to habitat degradation, biodiversity decline, and altered ecosystem function across a range of landscapes. These impacts are increasingly relevant under global change scenarios, where land intensification, vegetation clearance, and climate variability interact with dispersive soil behaviour in complex and often irreversible ways. Despite this, the ecological consequences of dispersive soils have historically received limited attention, especially compared with salinity, acidity, or contamination, leaving important knowledge and management gaps.

At the terrestrial scale, dispersive soils reduce vegetation cover, disrupt plant establishment, and fragment habitat. The formation of surface crusts impedes seed germination and root penetration, particularly in native grasslands, open woodlands, and dryland farming systems. Even when seeds do germinate, seedlings often fail to establish due to waterlogging at the surface or oxygen-deprived subsoils. As root growth is constrained, nutrient uptake is impaired, plant biomass declines, and soil organic matter inputs diminish. Over time, these processes generate barren or sparsely vegetated 'badland' patches that become focal points for erosion and degradation. In many landscapes, these patches expand, reducing the spatial continuity of habitat and lowering ecological resilience.

Soil biodiversity is also adversely affected. Dispersive soils tend to exhibit high pH, elevated exchangeable sodium, and low organic matter content, all factors that suppress microbial activity and limit habitat availability for invertebrates and soil fauna. Microbial biomass and enzyme activity are significantly lower in dispersive soils than in adjacent stable soils, indicating a decline in microbial function. The loss of pore space and aggregate integrity restricts microbial niche diversity, reducing functional redundancy and the soil's capacity for self-repair. These biological constraints further weaken plant–microbe interactions and nutrient cycling, exacerbating degradation.

Aquatic ecosystems bear the downstream consequences of dispersive soil erosion. Run-off from dispersive areas transports large volumes of fine clay particles into streams, wetlands, reservoirs, and estuaries. Unlike coarser sediments that settle quickly, dispersed clays remain suspended, increasing turbidity and reducing light penetration. This negatively affects aquatic vegetation, smothers benthic habitats, and disrupts the reproductive cycles of fish and macroinvertebrates. In coral reef and seagrass systems, such as those in northern Queensland, dispersive gully erosion has become a major contributor to sediment and nutrient loads, threatening both water quality and biodiversity. These effects are magnified in catchments with steep slopes, erodible soils, and limited ground cover.

In geomorphic terms, dispersion-driven erosion alters landscape connectivity and hydrology. Tunnel erosion, gully formation, and streambank collapse transform formerly productive or ecologically intact areas into degraded zones with accelerated water flow and sediment delivery. These features reduce infiltration, lower water tables, and can permanently alter catchment function. As channels deepen and widen, they disrupt ecological corridors and provide pathways for invasive species. Without active intervention, such landforms often worsen over time, creating positive feedback loops that are difficult and expensive to reverse.

8.6.1 Knowledge gaps

Despite growing recognition of these issues, important knowledge gaps persist. First, there are limited quantitative data on biodiversity loss associated specifically with dispersive soils. Although qualitative observations suggest declines in plant diversity, microbial richness, and soil fauna abundance, few studies have measured these changes systematically across gradients of dispersion severity. Furthermore, the thresholds at which dispersion begins to drive ecological change remain poorly defined. For instance, it is unclear at what level of sodicity or structural collapse vegetation fails to regenerate or microbial functions decline irreversibly.

Second, the potential for ecological recovery in dispersive soils is poorly understood. Although some observations suggest that native species (e.g. *Acacia* or *Eucalyptus*) can slowly improve subsoil structure through root activity and litter inputs, the mechanisms and timelines for such recovery have not been well documented. The role of natural succession in stabilising dispersive soils, as opposed to requiring human intervention, remains uncertain. Likewise, there is little understanding of how different ecological communities respond to dispersive stressors: whether some species exhibit tolerance, facilitation, or sensitivity that could be leveraged in restoration planning.

Third, there is a lack of integration between soil science and conservation biology in landscape planning and ecological restoration. Soil constraints such as dispersion are often overlooked in biodiversity or habitat mapping, leading to the establishment of restoration targets that may be unachievable without addressing underlying soil limitations. Similarly, many ecological restoration projects fail due to unrecognised soil structural issues that prevent vegetation establishment or hydrological function.

Finally, monitoring tools and indicators tailored to ecological impacts of dispersion are underdeveloped. Although erosion rates and turbidity are sometimes measured, more nuanced metrics, such as microbial biomass, infiltration capacity, or soil respiration, are rarely included in ecological assessments of dispersive areas. This limits the ability to track ecosystem response or detect early signs of improvement following remediation.

8.6.2 Future perspectives

Advancing understanding and management of the ecological impacts of dispersive soils requires a more integrated and interdisciplinary approach. Restoration ecology must incorporate soil structural diagnostics into project planning, including dispersion risk, ESP levels, and subsoil sodicity. This will help ensure that plantings are matched to soil capacity and that mechanical or chemical amelioration precedes revegetation where necessary. For instance, applying gypsum or compost to surface layers, or using deep-rooted species with known aggregation-enhancing traits, can prepare dispersive soils for successful re-establishment of native vegetation.

Geomorphic restoration techniques, such as reshaping gullies, applying vegetative stabilisers, or using coir logs and mulch, can significantly reduce sediment transport and improve ecological condition. Recent examples from the Great Barrier Reef catchments show that targeted gully remediation can reduce sediment delivery by over 90%, with co-benefits for downstream aquatic ecosystems. These strategies must be informed by detailed soil assessments and paired with hydrological modelling to ensure lasting success.

Ecological monitoring should be expanded to include soil-based indicators, such as microbial functional diversity, aggregate stability, or plant-available water capacity, alongside traditional biodiversity metrics. Coupling remote sensing (e.g. vegetation indices, thermal imaging) with *in situ* measurements can support large-scale surveillance and adaptive management. This is particularly important in rangelands, wetlands, and riparian corridors where dispersive processes may unfold over wide areas and long timescales.

On a policy level, dispersion should be recognised explicitly as a driver of land degradation, with implications for biodiversity offsets, habitat condition scoring, and land degradation neutrality targets under the United Nations Convention to Combat Desertification (UNCCD; Byron-Cox 2020). Environmental impact assessments should include dispersion risk screening, particularly in regions undergoing land use change or infrastructure development.

8.7 LIMITATIONS OF CURRENT AMELIORATION AND REHABILITATION PRACTICES

The management of dispersive soils has traditionally relied on chemical, physical, organic, and agronomic interventions to improve structural stability and productivity. Although some methods, such as calcium-based amendments and mechanical disruption, have delivered benefits in specific contexts, they often fall short in addressing subsurface dispersion or delivering long-term, cost-effective outcomes. Persistent barriers, including soil depth, amendment availability, scalability, and unintended environmental impacts, remain unresolved.

Chemical approaches, particularly gypsum, are the most widely used. Gypsum supplies soluble calcium to displace sodium on clay exchange sites, promoting flocculation and aggregation. However, its effectiveness is often restricted to surface layers, especially in dryland systems where limited water movement impedes calcium leaching into deeper soil horizons. In subsoils below 30–40 cm, gypsum generally fails to induce significant change unless mechanically incorporated, a costly and logistically difficult process. High application rates (>5 t/ha) further constrain adoption, and overuse may increase salinity risks or lead to sulfate leaching. Lime (calcium carbonate), although beneficial in acidic–dispersive soils, is less effective in neutral to alkaline conditions due to low solubility. Alternatives such as calcium chloride, phosphogypsum, elemental sulfur, and industrial by-products have shown promise in trials but are hindered by high costs, regulatory hurdles, or site-specific limitations.

Organic amendments like compost, manure, and crop residues can enhance aggregation and microbial activity, indirectly aiding calcium cycling. However, high application rates and deep placement are typically required to affect subsoil structure, challenges that are impractical in broadacre systems. Moreover, the persistence of improvements depends on amendment quality, microbial processes, and seasonal moisture. Without ongoing input, benefits often diminish.

Mechanical interventions, such as deep ripping, subsoiling, inversion tillage, and mounding, can break compacted layers and improve infiltration, but their effects are frequently short lived without concurrent chemical stabilisation. In dispersive soils, mechanical disturbance alone can exacerbate problems by increasing the risk of clay smearing, tunnel erosion, or rapid reconsolidation, particularly when done under moist conditions. These methods are also energy intensive and costly, limiting their use in extensive or low-input systems.

Vegetative strategies, such as cover crops, deep-rooted perennials, and salt-tolerant species, can help stabilise the surface, boost organic inputs, and reduce erosion. However, plant establishment is often poor in degraded dispersive soils, especially where compaction or sodicity persist. Without prior amelioration, vegetation alone is rarely sufficient to address subsoil constraints. Poor grazing management can further degrade vegetative cover and accelerate erosion.

Despite decades of effort, many dispersive soils remain degraded, unproductive, and prone to erosion. In regions of Australia, for instance, gypsum has been applied for over 30 years, yet large areas of dispersive subsoil remain uncultivable or marginally viable. These persistent issues underscore not only the technical limitations of current practices but also the need for integrated, innovative approaches that go beyond conventional strategies.

8.7.1 Knowledge gaps

A key limitation in current amelioration science is the lack of evidence regarding the combined and sequential application of treatments. Most studies focus on individual strategies in isolation (e.g. gypsum alone or deep ripping alone), yet field experience suggests that multiple soil constraints (e.g. sodicity, acidity, low organic matter, and compaction) often co-occur and interact. There is little guidance on how to optimise combinations; for example, when should gypsum be applied relative to organic matter? Is deep ripping more effective before or after amendment incorporation? What combinations provide the best cost–benefit outcomes for different soil types and rainfall zones?

In addition, there is a lack of long-term evaluation of residual effects and the sustainability of treatments. Although many interventions show benefits in the first one to two seasons, data on performance over 5 or 10 years is limited. Farmers and land managers require confidence that investment in subsoil amelioration will deliver lasting improvements, yet few economic models include timelines, maintenance costs, or thresholds for retreatment. The risk of diminishing returns, due to re-consolidation, declining biological activity, or insufficient rainfall for leaching, remains poorly quantified.

The economic and practical feasibility of large-scale amelioration also remains an unresolved issue. Many treatments are cost-effective only under high-input systems or where productivity gains can be guaranteed. For marginal soils, mixed enterprises, or low-rainfall zones, the return on investment is often uncertain. Moreover, treatments that offer broader public benefits, such as erosion control or reduced downstream sedimentation, are rarely incentivised under current land management policies. This results in an implementation gap, where technically effective practices are underutilised due to perceived or real economic barriers.

Lastly, monitoring and evaluation tools for assessing the success of amelioration are limited. Most programs rely on ESP or EC tests and yield responses, with few tools to track physical soil structure or microbial recovery in real time. The lack of accessible, low-cost methods for farmers to evaluate changes in aggregation, infiltration, or root penetration means that many adjustments are based on anecdotal evidence or delayed crop feedback.

8.7.2 Future research directions

To overcome these limitations, a new generation of integrated, scalable, and site-specific strategies is needed. One priority is the development of precision amelioration techniques. By leveraging digital soil maps, remote sensing, and on-the-go soil sensors, amendments can be applied variably within fields to target zones of highest dispersion risk. This reduces unnecessary input costs and ensures treatments are aligned with actual need. Platforms such as electromagnetic induction, proximal sensing, or yield mapping can be coupled with AI-based decision tools to guide treatment placement and monitor outcomes over time.

Innovations in soil conditioners, including synthetic polymers (e.g. polyacrylamides), engineered nanoparticles, or biologically derived binding agents, may offer novel avenues for improving soil structure with lower volumes and more persistent effects. For example, polyacrylamide treatments have been used in irrigation furrows to stabilise soil surfaces and reduce run-off; similar strategies may be adapted for dispersive landscapes, although concerns regarding environmental persistence must be addressed. Engineered mineral–organic composites and nanoclays are also being explored for their potential to alter cation exchange dynamics and enhance aggregation.

Hybrid intervention strategies that synchronise mechanical and chemical treatments may also enhance subsoil amelioration. Techniques such as controlled subsoil fracturing (e.g. explosive cracking, pneumatic injection) followed by targeted amendment delivery could increase treatment depth and effectiveness. Coupling amendment application with wetting–drying cycles may improve solubility and mobility, particularly for gypsum, in dense, dispersive subsoils.

Complementary biological approaches should be integrated into long-term rehabilitation programs. Deep-rooted cover crops, microbial inoculants, and rhizosphere-active species can provide low-cost, biologically driven improvements that stabilise earlier gains from physical or chemical interventions. These strategies may also extend the life of initial treatments by enhancing microbial activity and organic inputs over time.

Lastly, the development of decision-support tools that incorporate both agronomic and economic modelling is essential. Such tools should help land managers evaluate which combinations of interventions are most likely to succeed under their specific conditions and resource constraints. They should also support the development of incentive schemes, such as cost-sharing, stewardship payments, or ecosystem service credits, that align private decisions with public benefits.

8.8 RESEARCH NEEDS FOR DIFFERENT CLIMATE ZONES AND SOIL TYPES

Dispersive soils are not confined to a single region or soil type; they occur globally across diverse climatic zones and geological settings, from semi-arid plains and tropical uplands to temperate valleys and coastal flats. These soils may appear in Vertosols, Sodosols, Solonetzes, and silty loams, and they express their dispersive behaviour in context-specific ways that depend on rainfall regimes, temperature patterns, vegetation cover, and land use history. However, current knowledge of dispersive soil management is heavily concentrated in certain regions, most notably Australia, parts of India, and the US, whereas other areas remain under-researched.

Advancing global strategies to address dispersion requires tailoring research to the unique constraints, opportunities, and drivers present in each climate zone. The following subsections outline specific research needs for managing dispersive soils across arid/semi-arid, tropical/subtropical, and temperate/cool regions, with emphasis on both biophysical understanding and practical implementation.

8.8.1 Arid and semi-arid regions

In arid and semi-arid environments, where evapotranspiration exceeds precipitation, dispersion is often driven by a delicate balance between salinity and sodicity. These climate zones are common across Australia, North and East Africa, the Middle East, Central Asia, and south-western North America. In such settings, salts tend to accumulate at or near the soil surface due to limited leaching, creating transient flocculation. However, when rare rain events or irrigation occur, salts can be rapidly leached, triggering sudden dispersion. This pattern creates episodic instability, with high erosion potential during storm events.

Research in these regions must prioritise the dynamics of transient salinity and electrolyte depletion under low-rainfall conditions. Specifically, studies should aim to identify the tipping points at which soils shift from stable to dispersive behaviour during wetting events. Experimental data and modelling are needed to define safe operating envelopes for water inputs, whether rainfall or irrigation, that maintain flocculation while allowing leaching of harmful salts.

Given the scarcity of water, research should also focus on water-efficient amelioration strategies, such as shallow gypsum incorporation, banded applications, and the use of slow-release calcium sources that release ions over extended periods. Innovations in drip irrigation, polymer-enhanced wetting agents, or subsurface conditioners may offer new ways to control dispersion without relying on deep soil moisture penetration.

Additionally, the potential for *in situ* remediation using native mineral resources should be explored. In calcareous or gypsiferous landscapes, for instance, there may be opportunities to activate naturally occurring minerals through microbial, chemical, or physical means.

Wind erosion, which becomes more severe when crusts break down, is another under-researched topic in arid dispersive soils. Linking dust emission patterns to dispersion-prone areas under climate stress may be critical for public health and land management planning in drylands.

8.8.2 Tropical and subtropical regions

Tropical and subtropical climates are characterised by distinct wet–dry seasons, high rainfall intensities, and rapid biological turnover. Dispersive soils in these regions, found across South-east Asia, Central and South America, and parts of Africa, are particularly vulnerable to erosion during intense storms and often occur in degraded uplands, cleared forest margins, or shifting agricultural zones.

Here, the most pressing research need is the development of erosion control strategies adapted to dispersive behaviour. Conventional contouring or terracing may be insufficient on highly dispersive slopes. Instead, agroforestry systems, multispecies vegetative barriers, and hydrologically sensitive landscape zoning should be tested for their capacity to reduce surface run-off and improve soil structural stability.

Crop system research is also essential. In many tropical regions, transitions from flooded rice to upland cropping systems have inadvertently activated dispersion due to the loss of flocculating water regimes. Research must address the management of dispersive subsoils in such transitions and test cover crops, rotation strategies, and organic matter inputs that can buffer structural decline.

The high biomass potential of tropical systems offers unique opportunities for biological remediation, including the use of deep-rooted shrubs, leguminous trees, and green manure systems. These species may stabilise dispersive soils through organic inputs, root–microbe interactions, and enhanced infiltration. Ethnobotanical knowledge may further inform the selection of indigenous species suited for remediation.

In subtropical duplex soils with sandy topsoils overlying dispersive clays, perched water tables often exacerbate erosion. Targeted subsoil amelioration, such as deep gypsum injection or dual-rooting crop systems, may be required to manage lateral flow and restore infiltration pathways.

8.8.3 Temperate and cooler regions

In temperate climates, such as parts of North America, Europe, China, and Central Asia, dispersive soils may not manifest in dramatic gullying but still contribute to reduced crop productivity, subsoil constraints, and sedimentation. Seasonal wetting and drying, combined with freezing and thawing, further complicate the expression of dispersive behaviour.

Research in these zones should focus on detection and early diagnosis, because dispersive subsoils are often overlooked in temperate farming systems. Many farmers attribute poor drainage or root growth to compaction or acidity, rather than dispersion. Geophysical methods, proximal sensing, and high-resolution digital soil maps can help improve diagnosis.

In regions with freeze–thaw cycles, the role of seasonal processes in disrupting or enhancing soil structure is poorly understood. Although freezing may crack dense layers and improve infiltration in some cases, it may also fragment aggregates and accelerate structural decline in dispersive profiles. Long-term trials comparing different tillage and amendment strategies under freezing conditions are warranted.

Drainage design is another area of need. In intensively managed temperate systems, tile drains are common, but these often fail in dispersive subsoils due to clogging and slaking. Research should investigate how drainage methods interact with soil chemistry and whether controlled drainage systems can be modified to facilitate salt and sodium removal without promoting dispersion.

Vertosols in temperate regions may offer natural cracking pathways that can be exploited for deeper amelioration. Innovative strategies, such as amendment placement during dry periods or mechanical aids that align with natural shrink–swell cycles, may improve access to deep dispersive layers without excessive energy input.

8.8.4 Climate-smart research priorities across regions

Although dispersive behaviour is shaped by regional climate and geology, certain cross-cutting research themes apply globally. These include:

- developing region-specific ESP–EC–CROSS thresholds for identifying dispersion risk under varying rainfall regimes
- creating multiconstraint amelioration strategies that address sodicity, compaction, pH, and organic matter simultaneously
- integrating AI-driven mapping tools, real-time sensors, and machine learning into early warning systems for erosion and structural decline
- evaluating the residual and cumulative effects of amelioration under future climate scenarios, including elevated CO_2, drought, and storm intensity
- promoting South–South and North–South collaboration, allowing for knowledge exchange between countries with different experience levels and institutional capacity.

Ultimately, dispersive soil management cannot rely on universal recipes. A practice effective in semi-arid Vertosols may fail in tropical duplex soils or temperate silty loams. Developing regionally adapted strategies requires that soil–climate–system interactions be the basis for experimentation, innovation, and implementation. Establishing a network of long-term comparative sites, operating under shared measurement protocols, would allow benchmarking of treatments across agroecological zones. Coordination with global platforms such as the International Network of Salt-Affected Soils (INSAS) of the Food and Agriculture Organization of the United Nations (FAO) or its Global Soil Partnership (GSP) can support data harmonisation, capacity building, and investment alignment.

Innovation priorities should be aligned accordingly: in arid zones, focus on low-water-input amelioration and secondary sodicity prevention; in tropical zones, prioritise erosion buffering and biological stabilisation; and in temperate zones, concentrate on drainage integration and early detection. Across all regions, integrating digital soil diagnostics with landscape-level planning and stakeholder co-design will be essential for practical success.

8.9 HARNESSING AI, THE INTERNET OF THINGS, AND REMOTE SENSING TECHNOLOGIES FOR DISPERSIVE SOIL MANAGEMENT

The growing availability of digital tools, ranging from AI and machine learning to Internet of Things (IoT) sensor networks and satellite-based remote sensing, offers a transformative opportunity to improve the way dispersive soils are identified, monitored, and managed. These technologies can provide spatial and temporal insights that traditional soil assessments cannot, offering more timely, cost-effective, and scalable pathways to diagnose dispersion risk, guide targeted amelioration, and assess remediation outcomes.

Historically, the diagnosis of dispersive soils has relied on field-based tests (e.g. the Emerson crumb test), chemical thresholds (e.g. ESP or SAR), or visual assessment of erosion symptoms. Although informative, these methods are limited by their spatial resolution, labour intensity, and reliance on retrospective signs of degradation. In contrast, digital tools allow for proactive and predictive diagnostics, enabling more efficient interventions and long-term landscape monitoring.

8.9.1 Digital innovations for dispersive soil monitoring

Remote sensing technologies, including multispectral, hyperspectral, and thermal imaging from satellite and drone platforms, have become increasingly capable of detecting surface indicators of dispersion. For instance, persistently bare areas, rapid post-rainfall greenness decline, or high reflectance in specific bands may serve as proxies for crusting, sealing, or poor plant establishment on dispersive soils. When combined with digital elevation models and terrain analysis, remote sensing can also identify gully-prone or erosion-vulnerable areas linked to subsoil dispersion. Time-series satellite imagery, such as from Sentinel-2 or Landsat, can track seasonal vegetation recovery or sediment plumes in downstream water bodies, offering landscape-scale evidence of dispersion impacts.

The IoT provides further granularity through in-field sensor networks. Soil moisture, EC, turbidity, and temperature sensors can be installed at multiple depths and locations to monitor real-time changes in water movement, salinity levels, and infiltration rates, parameters critical to understanding dispersive behaviour. Turbidity sensors in run-off or drainage channels can detect episodic erosion events, whereas EC and sodium sensors can provide early warnings of sodicity build-up or electrolyte depletion. These sensor platforms, often connected via Long Range Wide Area Network (LoRaWAN), Narrowband Internet of Things (NB-IoT), or satellite uplinks, allow for continuous data transmission in remote or poorly connected areas.

AI and machine learning algorithms are increasingly used to process and integrate these diverse data streams. By training models on large soil datasets incorporating ESP, EC, clay content, yield maps, and historical erosion patterns, AI systems can identify patterns and generate predictive risk maps for dispersion. Machine learning techniques such as decision trees, support vector machines, or convolutional neural networks have been successfully used to predict sodicity and hydraulic behaviour across heterogeneous landscapes (Minasny *et al.* 2024; Yaghoubi *et al.* 2024). Importantly, these models can incorporate both static data (e.g. soil maps) and dynamic inputs (e.g. weather, irrigation schedules), allowing for adaptive management recommendations.

The development of decision-support platforms, which combine remote sensing, field data, and AI analytics, now enables land managers to visualise dispersion risk, track changes over time, and receive site-specific recommendations. For example, platforms like ConstraintID (developed in Australia) allow users to integrate soil test results with spatial data to prioritise amelioration interventions. Such systems are increasingly accessible via mobile apps, expanding their utility to farmers, extension officers, and planners alike.

8.9.2 Knowledge gaps

Despite their promise, digital technologies face several challenges in the context of dispersive soil management. First, the resolution and validation of remote sensing proxies for dispersion are still limited. Although spectral indices can detect bare soil, turbidity, or reduced vegetation, they cannot directly measure clay dispersion or aggregate stability. Calibration with field data is essential, yet many regions lack the necessary high-resolution soil surveys or long-term erosion datasets to support model training.

Second, there is a shortage of sensor technologies tailored specifically for dispersive soils. Many available soil sensors measure bulk EC or moisture but do not distinguish between salinity and sodicity, or between flocculated and dispersed conditions. There is a need to develop or adapt sensors that can detect key indicators of dispersive potential (e.g. real-time zeta potential, dispersion ratio, or turbidity on wetting) in practical, field-deployable formats.

Third, AI model performance depends on data quality and contextual relevance. Many soil prediction models are trained on limited datasets from specific regions (e.g. south-eastern Australia), and their generalisability to other climates, soil types, or management systems remains untested. In addition, the 'black box' nature of some AI algorithms may reduce trust among users, particularly if recommendations are not transparent or interpretable.

Fourth, connectivity and power constraints hinder deployment in remote or poorly serviced landscapes. IoT devices must be robust to extreme conditions, power efficient, and capable of transmitting data over long distances without reliance on cellular networks. Although satellite-enabled connectivity is improving, it remains costly and technically complex for widespread use in many low-income regions.

Finally, the translation of data into action remains a challenge. Many farmers and land managers lack the training, capacity, or incentive to interpret sensor outputs or use decision-support tools. Without appropriate extension services, co-design processes, and institutional support, digital innovations risk being underutilised or misapplied.

8.9.3 Future integration and implementation

To unlock the full potential of digital technologies for dispersive soil management, future efforts must focus on interoperability, accessibility, and scalability. Integrated platforms should be designed to bring together remote sensing, sensor data, and machine learning into user-friendly dashboards that prioritise key decision points, such as when to apply amendments, where erosion risk is highest, or whether dispersion risk has increased after a rainfall event.

Hybrid monitoring systems, combining satellite-based land surface analysis with IoT ground sensors, can offer complementary strengths. For instance, a regional dispersion alert could be triggered by satellite imagery showing post-storm vegetation loss, while in-field sensors validate the presence of turbidity or infiltration decline. Over time, these systems can build predictive models tailored to specific regions or farm types, allowing for proactive rather than reactive intervention.

New research should prioritise the development of low-cost, open-source sensor platforms designed for dispersion-relevant parameters. These could include portable soil turbidity meters, EC/Na ratio detectors, or Bluetooth-enabled tensiometers adapted for rugged field conditions. Such innovations will be particularly valuable in smallholder systems, where digital equity and affordability remain concerns.

Capacity building and participatory technology design must also be embedded in implementation efforts. Farmers, Indigenous communities, extension agents, and policy stakeholders should be engaged in defining priorities, validating outputs, and shaping digital interfaces. Doing so will improve adoption, trust, and contextual fit, while also enhancing scientific accuracy through ground-truthing and co-observation.

From a research standpoint, large-scale multiregion digital soil observatories could provide a shared infrastructure for advancing dispersive soil science. Such networks would collect harmonised soil–water–crop–climate data across climate zones, enabling comparative analysis, AI model training, and rapid dissemination of lessons learned.

In conclusion, digital technologies offer unprecedented tools for tackling the complexity and variability of dispersive soils. While technical, institutional, and economic barriers remain, the integration of AI, IoT, and remote sensing into dispersive soil diagnostics and management represents a critical frontier, one that can vastly improve the precision, efficiency, and sustainability of land restoration efforts under changing climatic and socioeconomic conditions.

8.10 INTEGRATING MULTIDISCIPLINARY APPROACHES: BRIDGING PHYSICS, CHEMISTRY, BIOLOGY, HYDROLOGY, AND SOCIOECONOMIC DIMENSIONS

Dispersive soils represent a convergence point for multiple soil degradation processes (physical, chemical, and biological), and their management requires an equally integrated response. Historically, different aspects of the problem have been addressed in disciplinary silos: soil chemists study ion exchange and salinity; physicists analyse permeability and dispersion kinetics; biologists examine microbial structure and plant–soil interactions; hydrologists model water flow and erosion risk; and engineers design infrastructure interventions. Yet dispersive behaviour is governed by intersecting processes that do not respect disciplinary boundaries. In addition, the success or failure of management practices is increasingly shaped by socioeconomic, policy, and institutional factors, which determine the feasibility, scalability, and adoption of technical innovations.

In this context, integrating disciplinary expertise is not merely beneficial, it is essential. Dispersive soils present one of the clearest cases where systemic thinking and collaborative research–practice models must replace reductionist or single-solution approaches. Building effective and sustainable soil interventions demands the coordination of science, technology, community knowledge, and policy in a shared framework.

8.10.1 Rationale for integration

The physical processes driving dispersion, such as aggregate disintegration, pore collapse, and erosion, are tightly coupled with chemical factors, including the balance of exchangeable cations, salinity, and pH. For example, the movement of water through dispersive profiles cannot be accurately predicted without accounting for both the chemical destabilisation of clays and the biological changes in organic matter content and microbial activity. Similarly, many of the chemical amendments used in dispersive soils (e.g. gypsum, lime) influence not only ion exchange equilibria but also microbial habitats, soil pH buffering, and biological decomposition rates.

Moreover, hydrological processes are deeply affected by changes in soil structure and chemical composition. Dispersion modifies infiltration pathways, increases run-off, and accelerates sediment transport, all of which feed back into nutrient cycling, vegetation establishment, and ecosystem services. Without a hydrological lens, the broader catchment-scale impacts of dispersive erosion and rehabilitation cannot be understood.

Socioeconomic dimensions add further complexity. The adoption of dispersive soil treatments, whether biological, chemical, or mechanical, is rarely determined by efficacy alone. Landholder knowledge, perceived risk, cost, access to information and materials, cultural practices, and institutional support all shape whether, how, and when interventions are applied. For example, a technically sound amendment strategy may fail in practice if it is not economically viable or if it is incompatible with local labour or machinery availability. The integration of disciplinary knowledge with practical realities is necessary not only for innovation but also to ensure that interventions are context sensitive, inclusive, and just.

8.10.2 Knowledge gaps in disciplinary coordination

Despite the clear need for integration, several structural and conceptual gaps persist. First, soil research remains highly fragmented, with few programs designed explicitly to foster collaboration across disciplines. Even when interdisciplinary projects are funded, disciplinary outputs often remain separate and uncoordinated, reducing the potential for synthesis and actionable solutions.

Second, there is limited formal engagement between soil science and socioeconomic research. Although agronomic and economic analyses are increasingly being incorporated into soil programs, the full range of human dimensions, including social acceptability, behavioural incentives, knowledge systems, and equity considerations, are rarely embedded. This weakens the capacity of soil science to engage with policy and limits the translation of innovations into practice.

Third, the absence of holistic performance metrics makes it difficult to compare or combine interventions. Although gypsum application may improve flocculation, it may not improve microbial function or plant resilience. A vegetative cover may reduce erosion but fail to ameliorate subsoil structure. Without frameworks that evaluate chemical, physical, biological, and economic outcomes together, it is challenging to assess trade-offs or synergies among different strategies.

In education and training, few programs prepare soil scientists, engineers, ecologists, and policy professionals to work collaboratively. The result is a shortage of professionals who are fluent across disciplines and able to bridge communication and methodological divides.

Finally, institutional structures and policy mechanisms are often siloed, with responsibilities for soil health divided among agricultural, environmental, water, and infrastructure agencies. This fragmented governance landscape limits coordinated investment and monitoring.

8.10.3 Pathways towards holistic soil solutions

Overcoming the barriers detailed above will require deliberate institutional, educational, and research strategies that promote systemic thinking and co-designed solutions. Several key directions can support this transition.

First, multidisciplinary soil restoration frameworks, such as the Soil Security Framework and One Health (Swan *et al.* 2024), should be operationalised to guide research, planning, and investment. These frameworks emphasise the interconnectedness of soil with water quality, ecosystem function, food production, and human wellbeing, offering an inclusive platform for collaboration.

Second, there is a need to support interdisciplinary case studies and demonstration projects that implement and monitor integrated soil rehabilitation strategies. For example, a regional pilot project may combine deep ripping with gypsum and compost, introduce rhizosphere-active species, and deploy IoT-enabled monitoring tools, while also engaging with local farmers, Indigenous knowledge holders, economists, and policymakers to design, monitor, and refine the intervention. Such initiatives can serve as real-world testbeds for understanding what integration looks like in practice.

Third, research funding schemes should prioritise cross-sector consortia that include biophysical scientists, economists, extension agents, land users, and institutional actors. Grant structures should incentivise shared problem definitions, co-publication across disciplines, and longitudinal impact assessments.

Fourth, the development of integrated decision-support tools is essential. These tools should allow users to assess multiple outcomes, such as yield gain, erosion reduction, economic return, and ecological benefit, from different intervention scenarios. Models should be designed for transparency and usability, with options to adjust for local conditions and constraints.

In education, soil science curricula should include systems thinking, stakeholder engagement, and exposure to diverse analytical methods. Training the next generation of soil professionals to operate in interdisciplinary teams will be essential to building future capacity.

Finally, policy and institutional reforms must be part of the integration agenda. National and regional soil strategies should establish cross-departmental coordination mechanisms to address soil health as a shared priority. Incentives for integrated land management, such as payments for ecosystem services, land stewardship programs, or carbon–biodiversity credits, can help align private action with public good.

8.11 FUTURE POLICY DIRECTIONS AND INTERNATIONAL COLLABORATION

Although the scientific understanding of dispersive soils has progressed significantly over the past three decades, policy frameworks at local, national, and international levels have yet to fully recognise or respond to the unique risks posed by dispersion. Unlike salinity, acidity, or erosion, which are often prominent in land degradation policy narratives, dispersion is frequently hidden or misclassified, despite being a major driver of soil structural decline and catchment-level sediment loss. Moreover, international mechanisms for funding, monitoring, and addressing land degradation rarely feature dispersion explicitly, resulting in underinvestment and fragmented response.

To effectively address the challenges posed by dispersive soils, a coordinated strategy is needed that bridges science, governance, and community practice. This requires embedding dispersion into soil policy instruments, regulatory frameworks, environmental assessments, and international collaboration platforms that link research with implementation across borders and sectors.

8.11.1 Policy recognition and regulatory gaps

Dispersive soils often fall between the regulatory definitions of salinity and erosion, despite contributing to both processes. For instance, current land classification systems may identify a soil as 'sodic' or 'saline' but not distinguish whether it is actively dispersive, intermittently dispersive, or structurally stable. This limits the ability of land use planners, regulators, and investors to make informed decisions about risk, land suitability, or remediation options.

In regions such as Australia, advances have been made in requiring dispersive soil management plans as part of major construction, mining, or agricultural development approvals. These plans typically require soil testing (e.g. ESP, dispersion tests), risk mapping, and site-specific remediation strategies. However, such policies are not yet widespread and, in many countries, dispersive soil risk is not included in environmental impact assessments or infrastructure planning processes, leading to recurrent failures of roads, dams, and embankments.

In agriculture, subsidy and extension schemes often prioritise salinity, pH correction, or erosion control without acknowledging the underlying role of dispersion in yield decline and run-off generation. Where dispersive soils are acknowledged, management recommendations may be outdated, overly general, or inaccessible to farmers without specialised support. This creates a policy–practice disconnect, where known solutions are not implemented due to lack of regulatory, financial, or technical support.

Furthermore, policies rarely integrate economic incentives for managing dispersion. Although some ecosystem service frameworks recognise erosion control or water quality improvement, the specific benefits of treating dispersive soils, such as reduced sediment transport or enhanced infiltration, are not quantified or monetised. This constrains investment in remediation and fails to reward proactive land management.

8.11.2 International collaboration and capacity building

Dispersive soils are a global phenomenon, with a documented presence in over 40 countries across all inhabited continents. Yet international efforts to map, monitor, and manage dispersive soils remain nascent. Initiatives such as the FAO's GSP and the INSAS provide promising platforms for collaborative action, but dispersion must be explicitly integrated into their scope, alongside salinity and sodicity.

International capacity building is essential. Many countries with extensive dispersive soils, particularly in Sub-Saharan Africa, South-east Asia, and Latin America, lack the institutional and technical resources to assess dispersion risk, develop site-specific interventions, or implement coordinated soil programs. Targeted knowledge exchange with countries that have advanced experience in dispersive soil management (e.g. Australia, India, the US) can accelerate learning and adaptation. This includes sharing diagnostic protocols, treatment strategies, monitoring technologies, and policy models through workshops, field trials, fellowships, and virtual platforms.

Joint research projects that compare treatment efficacy across soil types and climates can help generate globally relevant guidance. For instance, comparing bio-amendment strategies or digital soil diagnostics in India, Ethiopia, and Brazil would improve understanding of regional adaptation and innovation potential.

In addition, dispersion should be incorporated into international reporting mechanisms. The UNCCD's Land Degradation Neutrality framework and Sustainable Development Goal (SDG) Initiative 15.3 land degradation indicator, as well as the Global Assessment of Soil Pollution (UNCCD 2017), should all consider dispersion as a quantifiable form of degradation. Countries should be encouraged to report the area affected by dispersive soils, the interventions implemented, and progress in remediation.

8.11.3 Pathways for policy innovation and global integration

To support coherent and effective policy responses, several pathways can help mainstream dispersive soil management into national and international agendas.

First, national soil strategies should explicitly include dispersive soils within land classification, monitoring, and extension frameworks. This includes establishing diagnostic standards (e.g. minimum ESP–EC thresholds), updating soil survey protocols, and creating risk zoning maps that inform both land use planning and agricultural investment.

Second, regulatory instruments should mandate dispersion risk screening in infrastructure projects, land clearing applications, and irrigation expansion schemes. Engineering codes and environmental guidelines must require appropriate soil tests, design modifications, and mitigation measures to prevent failures and downstream impacts.

Third, economic policy instruments, such as cost-sharing grants, stewardship payments, or performance-based incentives, can support landholders who implement erosion control or dispersive soil amelioration. These should be integrated into broader climate adaptation, carbon farming, and biodiversity programs, reflecting the co-benefits of soil restoration for water security, carbon sequestration, and rural livelihoods.

Fourth, payment for ecosystem services schemes can be expanded to include sediment reduction and infiltration enhancement from dispersive soil management. This would require developing quantification tools to estimate avoided sediment loads, water retention benefits, or nutrient loss reduction, with payment linked to verified outcomes.

Fifth, multilateral partnerships between governments, research institutions, and non-governmental organisations can help fund landscape-scale programs focused on gully restoration, vegetation re-establishment, and precision amelioration. These initiatives should prioritise Indigenous land stewardship, gender equity, and local employment where possible.

Lastly, global research alliances, such as a global dispersive soils initiative, could coordinate data sharing, innovation testing, and policy harmonisation. Such a platform would support alignment across soil science, hydrology, engineering, economics, and policy disciplines, helping to elevate dispersive soils as a priority within the global soil health agenda.

8.12 LOOKING FORWARD

Dispersive soils represent one of the most persistent and under-recognised threats to global land productivity, water quality, and ecosystem function. Their presence undermines the sustainability of agriculture, accelerates erosion, degrades biodiversity, and compromises infrastructure, yet they continue to receive limited attention in mainstream soil policy and management frameworks. As this volume has demonstrated, dispersive soils are not merely a local agronomic constraint but are also a multidimensional, transboundary challenge, shaped by climatic variability, land use decisions, socioeconomic conditions, and global environmental change.

The chapters preceding this synthesis have collectively outlined the mechanistic complexity of dispersion, from the destabilising effects of sodium on clay particles to the cascading consequences for water infiltration, microbial function, and landscape structure. They have also illuminated critical knowledge gaps, including the limitations of current diagnostic tools, the need for regionally adapted amelioration strategies, and the underexplored roles of biological processes and emerging technologies. Most importantly, this work has highlighted a growing consensus: that no single discipline, technology, or intervention can resolve the issue of dispersive soils in isolation. To move forward, a new paradigm is required, one that recognises dispersive soil management as a systemic, interdisciplinary, and globally relevant challenge. This requires action on multiple, integrated fronts, as detailed below.

8.12.1 Reframing the science of dispersion

Scientific inquiry must evolve from characterising individual soil parameters to understanding the dynamic interactions among soil chemistry, physics, biology, and hydrology that drive dispersion across scales. This includes refining predictive indices such as CROSS and EC–ESP thresholds, expanding research into plant–microbe–mineral interactions, and integrating dispersion dynamics into hydrological and biogeochemical models. Soil function, rather than just soil condition, must become the new unit of analysis, focusing on how dispersive behaviour affects ecosystem services such as water regulation, carbon cycling, and biodiversity support.

8.12.2 Innovating context-specific solutions

Amelioration must shift from generic prescriptions to tailored, multiconstraint interventions that reflect local climate, soil type, resource availability, and landholder goals. This includes precision gypsum placement, polymer–biochar blends, microbial consortia, and root trait-driven plant selection. Innovations in digital soil sensing, AI-based diagnostics, and IoT-enabled monitoring will enhance both precision and accountability. However, innovation must be equitable and scalable, accessible not only to large-scale producers but also to smallholders and land managers in resource-limited settings.

8.12.3 Integrating social, economic, and institutional dimensions

Effective dispersive soil management requires more than technical tools; it demands an enabling environment that includes economic incentives, governance reforms, and capacity development. Landholders need support to adopt complex, sometimes costly interventions, particularly when benefits accrue over longer time frames or to downstream stakeholders. Policies must align private risk with public benefit through stewardship payments, erosion mitigation credits, or ecosystem service markets. Institutions should support coordination across agriculture, environment, water, and infrastructure sectors to avoid duplication and maximise collective impact.

8.12.4 Advancing global collaboration and policy alignment

Dispersive soils must be mainstreamed into global soil health frameworks, including the UNCCD's Land Degradation Neutrality targets, the FAO's GSP, and emerging climate–soil monitoring systems. International data repositories should include dispersion-specific layers and indicators, while funding programs must prioritise dispersion as a key driver of erosion, gully formation, and catchment degradation. Research partnerships between countries with advanced knowledge (e.g. Australia, India, the US) and those with emerging challenges (e.g. Sub-Saharan Africa, Central Asia, Latin America) will be essential to close knowledge gaps and build capacity.

A global dispersive soils observatory network, integrating long-term field sites, harmonised diagnostics, and standardised monitoring, could serve as a foundation for international benchmarking, knowledge exchange, and cross-regional learning.

8.12.5 Building a systems-based legacy

Ultimately, tackling the challenge of dispersive soils must become part of a broader movement to manage land as an interconnected system, where soil function underpins agricultural productivity, ecological resilience, and social wellbeing. This requires a shift in mindset: from reactive remediation to proactive resilience;

from fragmented interventions to integrated systems; and from soil as a background medium to soil as a central actor in global sustainability.

The task ahead is ambitious, but the foundation is already being laid. Across disciplines, sectors, and regions, momentum is building for a unified, science-based, and socially responsive approach to dispersive soil management. The challenge now is to sustain that momentum, bridge remaining silos, and convert knowledge into action. In doing so, dispersive soils, long treated as a stubborn liability, can be reimagined as a frontier of innovation, cooperation, and regeneration.

8.13 REFERENCES

Arnold JG, Srinivasan R, Muttiah RS, Williams JR (1998) Large area hydrologic modeling and assessment part I: Model development. *JAWRA Journal of the American Water Resources Association* **34**, 73–89. doi:10.1111/j.1752-1688.1998.tb05961.x

Byron-Cox R (2020) From desertification to land degradation neutrality: The UNCCD and the development of legal instruments for protection of soils. In 'Legal Instruments for Sustainable Soil Management in Africa'. (Eds H Yahyah, H Ginzky, E Kasimbazi, R Kibugi, OC Ruppel) pp. 1–13. (Springer International Publishing: Cham)

Carter MS, Tuttle MJ, Mancini JA, Martineau R, Hung C-S, Gupta MK (2023) Microbially induced calcium carbonate precipitation by *Sporosarcina pasteurii*: a case study in optimizing biological $CaCO_3$ precipitation. *Applied and Environmental Microbiology* **89**, e01794-22. doi:10.1128/aem.01794-22

Fang Y, Singh BP, Van Zwieten L, Collins D, Pitt W, Armstrong R, Tavakkoli E (2021) Additive effects of organic and inorganic amendments can significantly improve structural stability of a sodic dispersive subsoil. *Geoderma* **404**, 115281. doi:10.1016/j.geoderma.2021.115281

FAO (2015) 'Status of the world's soil resources: main report.' (Food and Agriculture Organization of the United Nations (FAO): Rome) Available at https://openknowledge.fao.org/server/api/core/bitstreams/6ec24d75-19bd-4f1f-b1c5-5becf50d0871/content [verified 30 July 2025].

Fu T, Saracho AC, Haigh SK (2023) Microbially induced carbonate precipitation (MICP) for soil strengthening: a comprehensive review. *Biogeotechnics* **1**, 100002. doi:10.1016/j.bgtech.2023.100002

Holzworth DP, Huth NI, deVoil PG, Zurcher EJ, Herrmann NI, McLean G, Chenu K, van Oosterom EJ, Snow V, Murphy C, *et al.* (2014) APSIM – evolution towards a new generation of agricultural systems simulation. *Environmental Modelling & Software* **62**, 327–350. doi:10.1016/j.envsoft.2014.07.009

Laflen JM, Lane LJ, Foster GR (1991) WEPP: A new generation of erosion prediction technology. *Journal of Soil and Water Conservation* **46**, 34–38. doi:10.1080/00224561.1991.12456572

McKenzie D, Bryce A (2023) Dispersive soil manual. Managing dispersive soils: practicalities and economics – Northern region. Grains Research and Development

Corporation. Available at https://grdc.com.au/resources-and-publications/all-publications/publications/2023/dispersive-soil-manual [verified 14 July 2025].

Minasny B, Bandai T, Ghezzehei TA, Huang Y-C, Ma Y, McBratney AB, Ng W, Norouzi S, Padarian J, Rudiyanto, Sharififar A, Styc Q, Widyastuti M (2024) Soil science-informed machine learning. *Geoderma* **452**, 117094. doi:10.1016/j.geoderma.2024.117094

Mueller CW, Baumert V, Carminati A, Germon A, Holz M, Kögel-Knabner I, Peth S, Schlüter S, Uteau D, Vetterlein D, Teixeira P, Vidal A (2024) From rhizosphere to detritusphere – soil structure formation driven by plant roots and the interactions with soil biota. *Soil Biology and Biochemistry* **193**, 109396. doi:10.1016/j.soilbio.2024.109396

Netrusov AI, Liyaskina EV, Kurgaeva IV, Liyaskina AU, Yang G, Revin VV (2023) Exopolysaccharides producing bacteria: a review. *Microorganisms* **11**.

Rengasamy P (2018) Irrigation water quality and soil structural stability: a perspective with some new insights. *Agronomy* **8**, 72. doi:10.3390/agronomy8050072

Rengasamy P, Marchuk A (2011) Cation ratio of soil structural stability (CROSS). *Australian Journal of Soil Research* **49**, 280–285.

Rengasamy P, Tavakkoli E, McDonald GK (2016) Exchangeable cations and clay dispersion: net dispersive charge, a new concept for dispersive soil. *European Journal of Soil Science* **67**, 659–665. doi:10.1111/ejss.12369

Rillig MC, Wright SF, Nichols KA, Schmidt WF, Torn MS (2001) Large contribution of arbuscular mycorrhizal fungi to soil carbon pools in tropical forest soils. *Plant and Soil* **233**, 167–177.

Rog I, van der Heijden MGA, Bender F, Boussageon R, Lambach A, Schlaeppi K, Bodenhausen N, Lutz S (2025) Mycorrhizal inoculation success depends on soil health and crop productivity. *FEMS Microbiology Letters* **372**, fnaf031. doi:10.1093/femsle/fnaf031

Sagar A, Rathore P, Ramteke PW, Ramakrishna W, Reddy MS, Pecoraro L (2021) Plant growth promoting rhizobacteria, arbuscular mycorrhizal fungi and their synergistic interactions to counteract the negative effects of saline soil on agriculture: key macromolecules and mechanisms. *Microorganisms* **9**, 1491. doi:10.3390/microorganisms9071491

Sahin U, Eroğlu S, Sahin F (2011) Microbial application with gypsum increases the saturated hydraulic conductivity of saline–sodic soils. *Applied Soil Ecology* **48**, 247–250. doi:10.1016/j.apsoil.2011.04.001

Sale P, Tavakkoli E, Armstrong R, Wilhelm N, Tang C, Desbiolles J, Malcolm B, O'Leary G, Dean G, Davenport D, Henty S, Hart M (2021) Chapter six. Ameliorating dense clay subsoils to increase the yield of rain-fed crops. *Advances in Agronomy* **165**, 249–300. doi:10.1016/bs.agron.2020.08.003

Shainberg I, Rhoades JD, Prather RJ (1981) Effect of low electrolyte concentration on clay dispersion and hydraulic conductivity of a sodic soil. *Soil Science Society of America Journal* **45**, 273–277. doi:10.2136/sssaj1981.03615995004500020009x

Swan T, McBratney A, Field D (2024) Linkages between Soil Security and One Health: implications for the 2030 Sustainable Development Goals. *Frontiers in Public Health* **12**, 1447663. doi:10.3389/fpubh.2024.1447663

Tavakkoli E, Pitt W, Armstrong D, Hilderbrand S, Fang Y, Uddin S, Armstrong RD (2022) Amelioration of hostile subsoils via incorporation of organic and inorganic amendments and subsequent changes in soil properties, crop water use and improved yield, in a medium rainfall zone of south-eastern Australia. (GRDC: Wagga Wagga). Available at https://grdc.com.au/resources-and-publications/grdc-update-papers/tab-content/grdc-update-papers/2021/02/amelioration-of-hostile-subsoils-via-incorporation-of-organic-and-inorganic-amendments-and-subsequent-changes-in-soil-properties,-crop-water-use-and-improved-yield,-in-a-medium-rainfall-zone-of-south-eastern-australia [verified 17 July 2025].

Trivedi P, Singh K, Pankaj U, Verma SK, Verma RK, Patra DD (2017) Effect of organic amendments and microbial application on sodic soil properties and growth of an aromatic crop. *Ecological Engineering* **102**, 127–136. doi:10.1016/j.ecoleng.2017.01.046

Uddin S, Rohan M, Weng ZH, Tahmasbian I, Fang Y, Hayden HL, Armstrong R, Tavakkoli E (2025) Enhancing root proliferation in an alkaline dispersive subsoil: a comparative study of organic and inorganic amendments with different amelioration mechanisms. *Journal of Soil Science and Plant Nutrition*. doi.org/10.1007/s42729-025-02602-w

UNCCD (2017) Scientific conceptual framework for land degradation neutrality. A report of the science–policy interface. Available at https://www.unccd.int/resources/reports/scientific-conceptual-framework-land-degradation-neutrality-report-science-policy [verified 30 July 2025].

Vakili AH, Salimi M, Keskin İ, Jamalimoghadam M (2024) A systematic review of strategies for identifying and stabilizing dispersive clay soils for sustainable infrastructure. *Soil & Tillage Research* **239**, 106036. doi:10.1016/j.still.2024.106036

Wang Y, Konstantinou C, Tang S, Chen H (2023) Applications of microbial-induced carbonate precipitation: a state-of-the-art review. *Biogeotechnics* **1**, 100008. doi:10.1016/j.bgtech.2023.100008

Wu W, Li A-D, He X-H, Ma R, Liu H-B, Lv J-K (2018) A comparison of support vector machines, artificial neural network and classification tree for identifying soil texture classes in southwest China. *Computers and Electronics in Agriculture* **144**, 86–93. doi:10.1016/j.compag.2017.11.037

Yaghoubi E, Yaghoubi E, Khamees A, Vakili AH (2024) A systematic review and meta-analysis of artificial neural network, machine learning, deep learning, and ensemble learning approaches in field of geotechnical engineering. *Neural Computing and Applications* **36**, 12655–12699. doi:10.1007/s00521-024-09893-7

Index

Page numbers in **bold** refer to figures.